The Princeton Review

Cracking the
Virginia SOL
EOC World History & Geography

by Dave Daniel

Random House, Inc.
New York

www.randomhouse.com/princetonreview

The Independent Education Consultants Association recognizes The Princeton Review as a valuable resource for high school and college students applying to college and graduate school.

Princeton Review, L.L.C.
2315 Broadway
New York, NY 10024

E-mail: comments@review.com

Published in the United States by Random House, Inc., New York.

ISBN 0-375-75566-7

Editor: Susanna Daniel

Manufactured in the United States of America

9 8 7 6 5 4 3 2 1

First Edition

Acknowledgments

For production and editing help, thanks to Lesly Atlas, Morgan Chase, Jeff Rubenstein, Stephanie Martin, Robert McCormack, Jennifer Arias, Raymond Asencio, Attrices Griffen and Evelin Sanchez-O'Hara.

Special thanks go to Alex Freer Balko for sharing her expertise on this project. A talented writer with the ability to see the big picture and an amazing eye for detail, Alex's work is always top-notch.

Contents

Chapter 1

Understanding the Basics

Thomas Jefferson Didn't Have to Take the SOLs

True enough. But, Thomas Jefferson and his contemporaries had their own challenges. One of your generation's challenges, unfortunately, is to maintain your sanity as you take an ever-increasing number of standardized examinations.

As you know, colleges like the University of Virginia and William & Mary have required students to take tests such as the ACT, SAT, and SAT IIs for admission for a long time. Now, the Commonwealth of Virginia has decided that you need to take more tests. This time a strong performance won't necessarily get you *into* college, but it will get you *out* of high school. In other words, whereas the SAT and ACT are college *entrance* exams, the SOLs are high school *exit* exams. If present trends continue, you'll soon have to take a test just to pass from one day to the next! *You passed the test? Good, you can move on to Thursday. You didn't pass the test? You have to stay in Wednesday.*

Okay, so maybe it's not *that* bad. But all these tests can seem a bit overwhelming. That's why we've written this book. We want to help you understand how the World History and Geography SOL tests are written and help you prepare for them so you won't have to take them more than once. This is the best guide you can use to prepare for the World History and Geography SOL exams.

The Mystery Exams

If you're reading this right now, chances are you already know something about the SOL assessments. You probably know that there's a lot of pressure on students, teachers, and schools to see that students do well, and that the tests will be a requirement for a diploma beginning with the graduating class of 2004. Don't worry. Whether you're a student, teacher, or parent, you've come to the right place for help.

Here is the story of the SOL exams in a nutshell: All students in Virginia are now required to take the SOL exams in all four basic subjects in grades three, five, and eight. The End-of-Course (EOC) exams in English, math, science, and history/social science are given to high school students at the completion of their classes beginning in grade 9. And starting with the class of 2004, students must pass a certain number of EOC SOL exams to earn credit for graduation. For each core course, there is a corresponding EOC exam that students are required to pass in order to receive a credit toward a high school diploma. This is the best guide you can use to prepare for the End-of-Course World History and Geography SOL exams. Our goal is to help you get in shape for this test.

How is this Book Going to Help Me?

This book has three parts. Part I, our strategy review, includes two chapters. Chapter 1 (this chapter) will help you understand how the test is organized and how to develop a study plan. Chapter 2 includes strategies to help you approach and conquer the multiple-choice questions on the exam.

> This book is broken into three parts: a strategy review, a content review, and practice tests. Read the parts in order to make the most of what we offer.

Part II, our content review, includes four chapters. Chapters 3 and 4 are subject reviews for the World History and Geography exam *until* AD 1000. Chapters 5 and 6 are subject reviews for the World History and Geography exam *since* AD 1000. Since this book helps you prepare for *both* tests, you need to focus on the two chapters that encompass the particular test that you are taking this time around.

Finally, part III contains two full-length practice exams, one for each of the courses for which you'll have to take a World History and Geography End-of-Course SOL examination.

Frequently Asked Questions

What Exactly Are the SOL Tests?

SOL stands for Standards of Learning. The SOL tests were first administered statewide in 1998. They are designed to find out whether students in Virginia are learning what they're supposed to be learning. What they're supposed to be learning is what's in the state curriculum for English, math, science, and history/social studies as outlined by the Virginia SOL.

Virginia's End-of-Course (high school) SOL tests:

- Will be mandatory for graduation—starting with the class of 2004.

- Measure what you have learned in key academic subjects.

- Test critical thinking skills—not just the ability to spit back facts.

- Assess how well individual schools are doing at teaching their students.

> If you want to know more about the Virginia state standards, try the Virginia Department of Education website: http://www.pen.k12.va.us/VDOE/Instruction/sol.html. Their SOL test info page is at: www.pen.k12.va.us/VDOE/NewHome/soltestinfo. It includes released and sample test items.

Courses and SOLs Required for a Diploma

Subject	# of courses needed for HS Graduation	# of passed EOC tests needed for Standard Diploma	# of passed EOC tests needed for Advanced Diploma
English	4	2	2
Mathematics	3	1	2
Laboratory Science	3	1	2
History and Social Sciences	3	1	2
Health & Physical Education	2	0	0
Fine Arts or Practical Arts	1	0	0
Electives	6	0	0
Student Selected Test	0	1	1
Totals*	22	6	9

*For an Advanced Diploma, you will need 24 credits to graduate.

Gulp! What Happens If I Don't Pass the SOL Tests?

If you are a member of the Class of 2004 or beyond, you *need* to pass six End-of-Course SOL exams (of the possible 12) to qualify for a standard diploma. If you pass enough of your classes, but fail to pass the necessary End-of-Course SOL exams, you will receive a **Certificate of Program Completion** in lieu of a high school diploma. That's the best possible scenario in the event the SOL exams are the only things hanging you up. Fortunately, you will be able to re-take any failed SOL test the next time it is offered. That means a student who fails a test in the spring of his or her junior year can re-take it after summer school, or in the fall, to meet the diploma requirement. Other than that, failure to satisfy the minimum diploma requirements will result in your participation in summer school.

Are There Any Other Awards That I Can Earn?

Why, yes, thanks for asking. For each type of diploma, there are also different levels of recognition that you can receive. If you complete a standard diploma (22 units, 6 verified credits) with an "A" average or better in your required course, you will receive a **Board of Education Seal** on your diploma. In addition, if you complete the requirements for an Advanced Studies diploma (24 courses and 9 passed SOL exams) with an overall "B" average or better AND you complete at least one Advanced Placement (AP) level course (or one college-level course for credit), you will receive a **Governor's Seal** on your diploma.

What If I Have a Disability?

Limited-English speaking students should consult http://www.pen.k12.va.us/VDOE/Assessment/LEPnrt.html for more information. Students with disabilities should go to http://www.pen.k12.va.us/VDOE/Assessment/SWDsol.html to see if they are exempt.

Currently, students with disabilities can get a special diploma that shows they've met their special-ed goals—if they are not taught the SOL curriculum. There is a new proposal that would allow students with disabilities to study the state curriculum and take SOL tests in early- and middle-school grades. In high school, they would take certain competency tests—yet to be determined—instead of the EOC English and math tests.

Students with learning disabilities and students who have learned English as a second language may be exempt from the regular EOC exams.

How Do the SOL Exams Compare With Other Standardized Tests?

Unlike standardized tests such as the SAT, students are *not* given percentile rankings that compare them with students across the country. Items do *not* progress from easy to hard to nearly impossible. The test is *not* timed, and it is *not* designed simply to measure your aptitude for learning.

How Are the EOC World History SOL Tests Organized?

Both of the World History and Geography exams contain multiple-choice questions, with four answer choices for each question. There will be no essays, no fill-in-blanks, no true-false questions. Only multiple-choice questions! We'll talk more about the structure of the exams in chapter 2.

How Do They Score This Thing?

Multiple-choice questions are easy to score. Scorers will add up the total number of correct answers. There's no guessing penalty, so don't leave any questions blank.

How Many Questions Do I Have to Answer Correctly?

For the World History to AD 1000 exam, you need to correctly answer approximately 33 out of 61 questions (54 percent) in order to pass, and about 55 out of 61 questions (90 percent) to get advanced recognition.

> Never, ever leave a question blank on the SOLs!

For the World History from AD 1000 to the Present exam, you need to correctly answer approximately 36 out of 63 questions (57 percent) to pass, and about 57 out of 63 questions (90 percent) to get an advanced score.

Think about this for a moment. It doesn't matter *which* questions you answer correctly, as long as you answer *enough* of them correctly. Any combination will do. This means that if certain areas of world history and geography are your strong suits, you can focus on those areas, and as long as the total number of correct answers adds up to enough points, you'll pass. Remember not to leave any questions blank, so you might even score a point or two by guessing. Of course, a strong score will reflect well on you, so you should try to obtain the highest score possible. But don't sacrifice a passing score by trying to do too much in too short amount of study time. If you only have a short amount of time to study before the exam, focus on your strong areas. Even if you only get 30 questions correct because you know the answer, if you guess on the other 31 and get one-fourth correct, you'll gain another 7 points, which will be enough to pass the exam.

Is This Test Going to Take a Long Time?

The great thing about both World History examinations is that they are untimed. In other words, you can take as much time as you need to finish. Since they both contain just over 70 multiple-choice questions, you're probably not going to need hours and hours to finish. You should take all the time you need in order to perform as well as you can. There's no need to miss a question simply because you felt rushed. If a particular question is giving you some problems and you don't want to think about it any longer, that's fine. Just move on and return to the hard question when you can give it a fresh look.

Do I Have to Review *Everything* from my World History and Geography Classes?

No, you don't have to review everything—but don't get too excited. If you want to do well, you will need to be familiar with many of the major topics, themes, people, and organizations discussed in your world history and geography courses. However, this doesn't mean that every topic is equally important. Instead, some topics and themes are tested much more frequently, and in much greater detail, than others. This book will help you decide where to focus your efforts.

If a question has you stumped, skip it and come back to it later.

First things first. Take the short practice quiz at the beginning of chapter 3 or chapter 5 (depending on which SOL exam you're studying for). Go ahead. There's no time like the present. When you're finished, score the quiz and compare your answers with the correct answers following the quiz. If you received a perfect score, go stand in front of a mirror and dance, dance, dance. If your performance was less than perfect, start determining your strengths and weaknesses, because you need to develop a study plan.

Excuse Me, Did You Say "Study Plan?"

Yes. A study plan. Don't worry! All this means is that you should figure out what you already know *before* you waste time reviewing stuff that you don't need to review. If you acquired this book at least a month before the actual exam, you will have plenty of time to review the strategies and take the practice exam. Run to the mirror. Remind yourself how cool you are. If you acquired this book only a couple of weeks before the exam, you'll still have enough time, but you might have to cancel a couple of dates or cut down on the channel surfing. Run to the mirror. Remind yourself how determined you are. If you acquired this book only a few days before the exam, we hope you did really well on the practice quiz and that you're doing well in your world history class in school. If not, focus your attention on chapter 2 and then on the areas of world history and geography that are tested most frequently. Then, of course, you'll need to take the full-length practice test at the end of the book. You won't have as much time to look over your mistakes and correct every error, but you'll be amazed how much you can improve your score if you stay focused. Run to the mirror. Congratulate yourself on how fortunate you are that you found the Princeton Review.

The best strategy is to start studying, using our techniques and information, well in advance of test day.

Time's a-wasting. Tell me more about this study plan!

There are two ways to study. You can review tons of material that may or may not be tested, or you can study the Princeton Review way. This book is designed to help you determine which themes, historical events, famous people, and geographical features are tested most frequently. The point is that you need to focus on things that are worth your time.

You probably already know that you need to come up with some method of organizing the stuff you have already learned. That's probably why you bought this book—you want to become test-wise! To do this, you need to concentrate on major themes, eliminate answer choices that are inconsistent with those themes, and avoid getting freaked out about the stuff that hardly ever gets tested. That's why you need to plan. There are three steps in developing a successful study plan.

1. Figure out what you know.

2. Figure out what you need to know.

3. Figure out how much time and effort it's going to take to make up the difference.

Step 1 is easy. You already accomplished it when you took the practice quiz at the beginning of chapter 3 or chapter 5 (depending on which exam you're taking). If you haven't taken the practice quiz yet, make sure you do soon.

Step 2 is also easy. We've done most of the work for you. The stuff you need to know is listed in chapters 3 through 6. The hard part is step 3. You need to get a general idea of how much stuff in chapters 3 through 6 is totally new to you. If there's a lot, you need to develop a study plan that will give you enough time to not only read the information a couple of times but also organize it in your head. For example, if you say "Confucius" when somebody sneezes and think that "Mandela" is a new restaurant, you better plan on spending a lot of time reading the content review chapters. Why? Because Confucius and Mandela are part of the regular cast of characters that appears on each of the exams (Confucius on the World History to AD 1000 exam, Mandela on the World History from AD 1000 to the Present exam). How do you know which people and events are frequently tested? We'll tell you in the content review chapters! But first, you have to commit to giving yourself enough time to read them. That's why step 3 is so difficult. You need to schedule enough time to get the job done, and get it done right.

There's never enough time!

We know you're busy. You've got homework from school. You've got activities, family obligations, and maybe a part-time job. And now you're supposed to study for this exam on top of it all? Okay, okay. You have our sympathy. But you still have to take this test. And believe us, you only want to take it once. It makes sense to give this exam the priority it deserves. So here's what you do: Go get a calendar. Put a big circle around the day of the actual

> We've done a lot of the work for you by figuring out which areas of information will probably be tested. The least you can do is formulate an effective study plan.

test (we hope it's not tomorrow!). Next, figure out when you're going to take the practice exam in the back of the book. Try to give yourself at least a week before the real exam since you'll need time after the practice exam to learn from your mistakes. Now, depending on how well you did on the practice quiz and how well you're doing in your world history course in school, you also need to schedule some study time.

Study time includes three components:

1. Time to read chapters 1 and 2, which you should try to do within the next couple of days.

2. Time to study the review sections in chapters 3 and 4, or 5 and 6 (depending on which of the two World History exams you are taking).

3. Time to review your mistakes on the practice exam and learn from those mistakes. If you can get to the point where you can explain why the right answer is right and why the wrong answer is wrong, you've succeeded.

Preparing for this exam doesn't have to be overwhelming. Just stick to an orderly schedule.

How Do I Know if my Study Plan is Working?

Let's say you incorrectly answer a multiple-choice question about the geography of Japan. Your incorrect answer usually can be blamed on one of two mistakes: Either you were *clueless* or you were *careless*. If you were clueless on a question (meaning you remembered nothing about Japan or its geography), then you need to go back to the content review in chapters 3 through 6 and study the information about Japan.

There's no excuse for missing questions due to careless reading. Take a deep breath, and slow down.

If, however, you remember all the major stuff about the geography of Japan and still missed the question, then you were probably careless. You either misread the question, misread one of the answer choices, or just plain got tricked. In this case, you don't need to review the geography of Japan nearly as much as you need to review the test strategies in chapter 2. You might also need to slow down. Since you'll have an unlimited amount of time to complete the exam, there's no excuse for missing questions due to carelessness. Finally, sometimes you think you know about a certain topic, but in fact you don't. In other words, you might have thought Japan was a geographically large nation attached to mainland Asia, when, in fact, it's a geographically small, mountainous island nation. When you find yourself particularly surprised with an answer choice, a little alarm should go off in your head: *Warning! Bad information is floating around in my head. It must be eliminated and replaced with good information immediately!*

Got it. What Now?

Read chapter 2 first and then the content review chapters in part II of this book. Then, develop your study plan accordingly, depending on two things: how much you need to improve (a little or a lot) and how much time you have (a little or a lot). Then, follow your study plan. When you've finished this study plan, you will also have finished this book, and you'll be ready for the End-of-Course SOL World History and Geography exams!

What Exactly *Is* the Princeton Review, Anyway?

The Princeton Review is the nation's leader in test preparation. For nearly two decades, we've been helping hundreds of thousands of high school students prepare for the ACT, SAT, SAT IIs, and AP exams, along with many other state tests.

We want to help you become test-wise so that you won't be surprised on test day. We've scoured all the available information related to the SOLs. First, we reviewed the standards of learning that the tests measure and evaluated every sample question released by the Virginia Department of Education. Then, we applied our extensive knowledge of test development and world history and geography to develop test-taking techniques to help you score higher. Finally, we developed practice tests that look like the real End-of-Course SOLs and included them in this book. In short, we've done a whole world of work for you. The best preparation for the EOC SOL is right here in your hands. Read it. Study it. Ace the test. And then, for crying out loud, move on with your life.

Chapter 2

Structure and Strategies

What's on this Test, Anyway?

The World History to AD 1000 & World Geography exam has a total of 71 questions. The World History from AD 1000 to the Present & World Geography exam has a total of 73 questions. On both tests, there are ten questions known as "field-test items." These questions do not count toward your score. The test writers use them for their own research. Unfortunately, you won't know which questions are the field-test items and which are the questions that count, so you'll have to answer all the questions as if they count toward your final score.

Study efficiently by concentrating on the areas of information that you know will be tested.

Fortunately, we do know which learning standard categories will be tested and how frequently each will be tested. This is extremely important information since it will enable you to focus on the areas of world history and geography that are worth your time. As you can see in the charts below, the test writers and the Virginia Department of Education (VDOE) believe that some aspects of world history are more important than others.

World History to AD 1000 & Geography Exam

Test categories	Number of questions
Ancient civilizations	7
Greece and Rome	11
Middle East, Russia, and early medieval Europe	9
Asia, Africa, and the Americas	8
History skills	8
Geography skills	6
Geography knowledge and concepts	12
Total questions scored	61
Field-test items	10
Total questions on the exam	71

World History from AD 1000 to the Present & Geography Exam

Test category	Number of questions
Late medieval Europe: AD 1000 through the Reformation	8
Age of Discovery	8
Sixteenth through nineteenth centuries: Enlightenment, Absolutism, Reason, and Industrial Revolution	12
Twentieth-century world conflicts	10
History skills	7
Geography skills	6
Geography knowledge and concepts	12
Total questions scored	63
Field-test items	10
Total questions on the exam	73

Tell Me More About This Test

Each multiple-choice question on the exam has four answer choices. Odd-numbered questions have answer choices **A**, **B**, **C**, and **D**, while even-numbered questions have answer choices **F**, **G**, **H**, and **J**. The test writers do this to help you, not to confuse you. Sometimes students skip a question in the test but forget to skip a space on the answer sheet. By using different letters for every other question, the test writers help to prevent that kind of mistake from occurring.

Here are some other things to keep in mind:

- Answer choices such as "none of the above," "all of the above," and "not here" will not be used.

- Numerical answer choices, such as dates, will be arranged in ascending order. In other words, the first answer choices will be the smallest number or least recent date and the last will be the largest number or most recent date.

- Negative words (such as *not*, *least*, or *except*) will be emphasized by *italics*, **boldface type,** underlining, or CAPITAL LETTERS. Don't forget to read the questions carefully!

> If you skip a question on the exam, don't forget to skip a space on the answer sheet, too.

Process of Elimination

Each of the World History and Geography SOL exams has just over 70 multiple-choice questions. On some, you'll know the correct answer immediately, especially after you prepare with the help of this book. On others, you'll be uncertain and you'll have to make your best guess. On a few, you might have no idea at all. The thing to keep in mind is that since there isn't a guessing penalty, there is absolutely no reason to leave any questions blank. If you aren't certain of a correct answer, eliminate wrong answer choices first. If you are still uncertain of the correct answer after you've eliminated as many wrong answer choices as you can, guess from among the remaining answer choices.

The questions on the exam will be structured in many different ways. Some of the questions will be very straightforward—they might ask you about one characteristic of a person, such as Alexander the Great or Adolf Hitler, or of an event, such as the fall of Rome or the Salt March in India. Other questions will include descriptions of ideas and concepts, such as capitalism or divine right theory. Still other questions will include charts, graphs, cartoons, lists of events, or even quotations. Some questions will have one-word answer choices.

Others will have longer, descriptive phrases in the answer choices. Keep in mind that some of the techniques described below only apply in certain situations. Some of them are best used when there are long, descriptive phrases in the answer choices, and others are best used when the questions and answer choices are very straightforward.

We'll tell you in which situations you should use each technique, so please read carefully. This is definitely not a chapter you want to skim. You need to understand the techniques—and the circumstances in which you should use them—in order to benefit from this chapter.

Did You Hear the One About the Sculptor and the Elephant?

When you use Process of Elimination techniques, you don't necessarily need to know the answer to a question to get it right.

A popular children's joke goes something like this: How did the sculptor make an elephant out of a block of marble? He chopped away everything that didn't look like an elephant. It's the same deal on the test—no joke. It doesn't matter if you get the question right because you know what the correct answer looks like or because you know what the correct doesn't look like. In both cases, you'll get credit for answering correctly.

On the pages that follow, you will learn how to approach the End-of-Course World History and Geography SOL exams from two different angles. The first group of techniques will help you anticipate the correct answer so that when you see it, you'll know it's the right one. The second group of techniques will help you eliminate incorrect answer choices when you're not certain of the right answer.

Group 1: Spotting the Elephant

Déjà vu

It's easy to find the right answer when you've seen the information before. All of the commonly tested standards of learning are covered in the content review chapters in this book. If you read those chapters and organize the information in your head, many of the correct answer choices will pop right out at you because you will have seen them before.

Open your mind

If you don't understand POE after reading this section, ask a friend to explain it. POE is guaranteed to help you score higher on the exam!

When a question asks about the impact of a particular event, open your mind to the possibilities. Of course, you want to stay focused on the ideas and concepts that are important to the test writers, but you want to open your mind enough to see *all* of those ideas and concepts. For example, if a question asks you about things that impacted the ancient culture of Japan, one thing you should think about immediately is religion, since the dominant religion of a region virtually always impacts the region's culture. In the content review chapters of this book, you will be reminded that the dominant Japanese religions historically have been Buddhism and Shintoism.

The religions of a region are important, the major political and social developments in a region's history are important, and a region's geography is important—just to name a few factors. When answering a question about influential factors in a region's development, think about everything you know to be true about the region, as well as everything you know is *not* true.

The hip bone's connected to the leg bone

As you take the practice exams in this book, it's a good idea to pay attention to words that the test writers consistently group together. For example, *subsistence farming* often is linked with *traditional societies*, while *commercial farming* and *Green Revolution* often are linked with *modernization* or *westernization*. As such, if you see the term *subsistence farming* in a question, you should immediately look for the terms *traditional society* or *traditional culture* in the answer choices. If those exact words aren't in the answer choices, then pick the answer that *refers* to traditional societies.

Word association works for people as well. If you see *Karl Marx* in a question, the correct answer will likely refer to *communism* or describe some aspect of communism. If you see *Adam Smith*, the correct answer will probably refer to *laissez-faire economics* or *free-market system*.

Don't underestimate the importance of this concept. When you learn about a new person or a new event, don't simply memorize a definition. Link the definition to other words, people, and events. This strategy will help you learn larger concepts, and understanding larger concepts will help you spot the correct answers immediately.

Chapters 3 through 6 will provide you with the terms, people, and concepts that you need to link together, but it's up to you to do the linking. **Remember:** You can't take this book with you into the exam. You have to read it to make these connections.

Make sure you know what you know

The best way to find the right answer is to make sure you understand what the question is asking in the first place. After you take the practice tests, review the questions that you missed. If it turns out that you missed a few questions because you misread a question, you need to slow down a bit and read more carefully.

Read carefully. There's no sense losing points because you don't understand a question.

If you don't know the answer to a question even after rereading it carefully, you can still use some of the Process of Elimination techniques. Bad things happen when you *think* you know something that in fact you don't. Try to correct any misinformation that's floating around in your head. It's better to say, "I don't know for sure," than to say, "Oh yeah, I know this," when in fact you don't. When you're conscious of the fact that you don't know something for sure, you'll be more careful when evaluating the answer choices.

In the end, if you aren't able to find the correct answer to a question based on your knowledge and careful reading, then eliminate incorrect answer choices. This is what the next group of techniques is all about.

Group 2: 13 Ways to See Everything that Doesn't Look Like an Elephant

1. Just because you don't know the answer doesn't mean you don't know anything.

Okay, so you read a question and you don't know the correct answer. Let's say that the question refers to the Hwang Ho River valley. You can't remember anything about the Hwang Ho River valley. You're not even sure where it is, or even what it's near! Pull yourself together and ask yourself, "Okay . . . what *do* I know?"

Well, for one thing, you know it's a river valley! Therefore, whatever the answer is, it has to be something that is true of a river valley. Got it? Look at the following sample question.

Example 1

   A study of the ancient Liver River valley would be most important in understanding

   A why ancient empires were built from nomadic civilizations

   B the role of geography in the development of early civilizations

   C the rise of the onion as a major agricultural export

   D the expansion of the Roman Empire

You've probably never even heard of the Liver River . . . at least we hope you haven't because it doesn't exist. But that's the whole point of this little demonstration. Sometimes it doesn't matter if it's the ancient Liver River, the ancient Shiver River, or the ancient Joan Rivers— you should still keep in mind that it's a river! What's more, the test writers threw in the word "ancient," which means you're not thinking about just any river valley—you're thinking about a river valley that existed a long, long time ago.

Okay, let's evaluate the answer choices. We have the Liver River. We know it's ancient. We want to know what we'd learn if we studied the ancient Liver River. Here we go:

A This answer choice has the word "ancient" in it, but it doesn't make sense. To be sure, we don't know anything about the Liver River, but nomadic civilizations are civilizations that go from place to place in search of food or water for themselves and their herds. If they stumble across a river valley, there's little reason to remain nomadic. With a river nearby, they could stay in one place. Why in the world would *any* river valley be the reason behind a nomadic lifestyle?

B A river valley is definitely a geographic feature. And the ancient Liver River might have been at the heart of an early civilization. In fact, all of the earliest civilizations formed around rivers. So, even though we don't know the particulars of the Liver River, this answer choice is consistent with everything that we know about river valleys in general.

C Liver might indeed remind you of onions, depending on what you're forced to eat at dinnertime, and agricultural development works nicely with river valleys, so you might hang on to this answer choice for a bit. But once you compare it to answer choice **B**, and once you think about the significance of onions, this answer choice is less attractive. Why would the test writers waste a multiple-choice question on something as trivial as the development of onion farming? Onion development certainly has had an impact on some people's lives—onion farmers, for example. But is it truly essential to an understanding of world history and geography? Rice is essential because rice is a staple grain for billions of people and enormously affects the economies of Southeast Asia. But onions?

D River valleys definitely have been central geographic features of many empires, and the Roman Empire is an ancient empire, so you might think this is the correct answer. However, if you remember that the Roman Empire used the Mediterranean Sea as its central waterway, you might be skeptical of this answer choice. If you really open your mind, you also might recall that the Romans are revered for their road building, which means that they probably weren't highly reliant on river travel.

The correct answer, of course, is **B**.

The point is simple but important. The words in the questions themselves often provide clues to the right answer, even if you don't fully understand the question. When the questions are extremely straightforward with no descriptive words, it's difficult to use this particular technique. But when the questions include descriptive words, don't let them just sit there—build from them! Read every part of each question and every part of each answer choice, looking for clues. Then open your mind, arm yourself with the following elimination strategies, and find the right answer!

2. If it never happened, get rid of it.

Let's say you encounter a question about Japan. Let's also say that you don't entirely understand the question and you're uncertain of the answer. You should still read the answer choices to eliminate ones that you know are incorrect.

> If you're not sure of an answer, read each choice carefully, then eliminate those that just don't make sense.

If a question asks you to identify something about Japan, you can eliminate answer choices that refer to Japan's alliance with the United States during World War II. Why? Because Japan and the United States fought *against* each other during World War II.

Regardless of the question, always eliminate answer choices that describe events that never occurred.

3. But what if I don't know if the stuff in the answer choices actually occurred?

When you just don't know if the stuff in the answer choices actually occurred, don't give up. Instead, ask yourself this question for each answer choice: Even if it *did* occur, would it matter? If the answer is *no*, then you need to get rid of that answer choice. This technique is most helpful when a question asks you to determine the cause of something. Let's look at an example.

Example 2

The ethnocentric attitudes of various Chinese emperors were most likely caused by

   F the cultural isolation of China

   G the failure of other nations to become interested in China

   H the interest of Chinese scholars in other civilizations

   J the great cultural diversity within China's borders

This question might be difficult if you don't know what *ethnocentric* means. On the other hand, by the time you finish this book, you'll know what it means. Once you understand that *ethnocentric* means thinking that your own group is superior to other groups, you still have to evaluate the answer choices, and you might remain uncertain of whether the stuff in the answer choices actually occurred. Therefore, you need to ask yourself: Even if this stuff happened, would it be a reason behind ethnocentric attitudes in China?

Here we go again:

F Does it make sense that cultural isolation would have influenced ethno-centric attitudes? Sure, it does. When you isolate yourself from others, it's easy to think that you're better than others are because you never have to prove yourself.

G Even if other nations weren't interested in China, would this cause ethno-centric attitudes in China? No. Other people's interests would affect their own attitudes, not Chinese attitudes—at least not directly. (By the way, this answer choice doesn't describe something that actually happened in the first place, since other nations *were* interested in China.)

H Even if it's true that Chinese scholars were interested in other civiliza-tions, would this be the reason behind ethnocentric attitudes in China? No. Interest in others usually leads to mutual understanding, thereby decreasing, not increasing, ethnocentrism.

J Even if it's true that there is great cultural diversity within China's borders (and it is true), would this be the cause of ethnocentric atti-tudes in China? No. Diversity usually leads to mutual understanding. Isolation, on the other hand, usually leads to a singular point of view.

> Focus on answer choices that are *logically* related to the information in the question, even if you don't know.

The correct answer, of course, is **F**.

4. I like this answer because it's difficult to argue with.

Sometimes the test writers will give you a gift. It's like they're checking to make sure that you're paying attention. This happens when they give you an answer choice with which you just can't argue. If an answer has just got to be true, then it is. If a question refers to a war—*any* war—then it just makes sense that "people's lives are disrupted." If a question refers to a successful independence movement—*any* successful independence movement—then it just makes sense that "the colonial power weakened." If a question refers to a nation—*any* nation—with mountain ranges and deserts, then it just makes sense that "transportation is difficult." Is this too easy? Yeah, it is. But that's the point. There's no need to make questions harder than they are. Not every question has an answer choice with which you just can't argue, but when you come across one, pick it. And if you're guessing, at the very least choose the answer that is the *hardest* to argue with, even if you can argue with it a little bit.

5. Broad language overshadows other choices.

When evaluating answer choices, be attracted to answer choices that use broad, general language, especially when there are other answer choices that seem very narrow. Look at the following example:

Example 3

One way in which the Vedas, the Bible, and the Koran are similar is that these religious texts

A provide guidelines to govern the behavior of believers

B deny the existence of a supreme being who rules the universe

C encourage strife between segments of believers

D support political rebellions to overthrow existing governments

Even if you don't remember anything about the Vedas, the Bible, or the Koran, you can still answer this question correctly since the question tells you that these texts are religious texts. There are a couple of ways to approach the answer choices. First, you can focus on answer choices that are difficult to argue with, given that these are religious texts. If you find this unhelpful, focus on answer choices that use general language as opposed to specific language, especially if the general language also makes good sense. Answer choice **B** doesn't make much sense since these are religious texts.

Now, look at the other three carefully. If you like answer choice **C**, then you also have to like answer choice **A** since "strife between segments of believers" would be a type of "behavior of believers." In other words, if answer choice **C** is true (and it's not), then so is answer choice **A** (but not vice-versa). What's more, "political rebellions" in answer choice **D** is also a kind of behavior of believers, so if answer choice **D** is true, so is answer choice **A** (but not vice versa). Therefore, **C** and **D** can't be the answers because if either one of them was correct, answer choice **A** would also be correct. And we can't have more than one correct answer. So, what can we conclude? Answer choice **A**, the answer choice that uses the most general language, is the right answer. Always be attracted to answer choices with general language (of course, in situations with one-word answer choices, this technique won't apply).

> If all else fails, choose the answer choice that uses the most general language.

6. Don't be so extreme.

When evaluating answer choices, be alert for too-specific language *and* for language that is extreme. The test writers sometimes throw in an answer choice that is stated way too strongly. For example, "Economic opportunities improved for *everyone*"; "The example shows that *all* nations are committed to the *elimination* of nuclear weapons"; "Frustration over taxation was the *only* motivation behind the revolt." Correct answer choices usually are worded with much safer, more middle-of-the-road language. For example, "Economic opportunities improved for *many* people"; "The example shows that *some* nations are committed to a *reduction* in the production of nuclear arms"; "Frustration over taxation was *one of the main* motivations behind the revolt." This is not to say that every single time you see an extreme word you should eliminate it automatically—that action itself would be too extreme. The point is that you need to be suspicious of extreme language, especially when a safer answer choice exists. However, if a question asks about a totalitarian regime, for example, perhaps some of that regime's actions or policies were extreme, in which case extreme language might be justified.

Look at the following example:

Example 4

> **A study of Japan's economy since World War II would lead to the conclusion that**
>
> **F** Government support of technological advances can improve a nation's economic position.
>
> **G** Imperialism is necessary for the economic development of a nation.
>
> **H** A communist system leads to economic prosperity.
>
> **J** The feudal system is more economically productive than the market system.

> Trust yourself. If an answer choice doesn't seem to make sense, it probably doesn't. Eliminate it.

Which answer choice uses the safest language? Answer choice **F**, of course. And once you keep in mind all of the other techniques you've learned, it should be clear that **G**, **H**, and **J** are just plain *wrong*.

7. Hey buddy, do you have the time?

One of the most effective ways to eliminate wrong answer choices is to use your knowledge of *time periods*. We're not suggesting that you sit around and memorize dates. We're suggesting that you know the major time periods of world history and understand generally what was going on during each of these time periods. The information in the content review chapters will help a lot.

For example, if you're not sure about the details of the event described in a question, you might remember that it occurred during the Middle Ages. Then, answer choices that relate to *feudalism* might be correct while answer choices that relate to *socialism* or *humanism* are most certainly wrong, since neither of these concepts developed until centuries later.

8. Where am I, anyway?

If you can't remember specifics about a given historical figure or event, choose the answer choice that seems most closely related in general terms.

Just as a firm grasp of time periods can help you eliminate incorrect answer choices, so can a firm grasp of geographical locations. Keep a world map with this book. The content review chapters will help you organize all the stuff you need to know about geography. The thing to keep in mind, though, is that even if a question doesn't ask specifically about the *geography* of a region, you still need to understand the general characteristics of that region. Try to eliminate answer choices that are inconsistent with the developments of the region. If a question refers to Latin America, get rid of answer choices that mention Buddhism. If a question refers to a country in Africa, get rid of answer choices that refer to Divine Right Theory. You get the idea.

9. Wish lists

Sometimes you'll get a list of three names, events, locations, or other things. The question will ask you, "What do these three items have in common?" Students are often intimidated by lists because they think they need to know about three things instead of just one. Don't be intimidated! These questions are actually gifts because they provide more than one opportunity to find the correct answer. Here's the deal: You don't necessarily have to know about all three things in a list to answer the question correctly. In fact, you usually only have to know about one of them, sometimes two. If you are familiar with only one of the items in the list, start eliminating answer choices that are not consistent with that one item. If you have more than one answer choice remaining, evaluate the remaining answer choices in light of the other Process of Elimination techniques you've learned.

10. Major means MAJOR.

Many test questions include clues. If a question asks about a MAJOR event, you'll know to look for a MAJOR answer choice, even if you don't know the correct answer.

The test writers often will use the words *major* or *significant* or *important* in a question: "A *major* problem currently facing the Republic of South Africa is the . . ." or, "A *significant* goal of the Meiji government in Japan was to . . . ." Even if you don't remember much about South Africa or the Meiji government, you still want to focus on answer choices that describe something MAJOR, something HUGE, something that would have a BIG IMPACT. If *all* the answer choices describe big, important stuff, then you'll have to use other Process of Elimination techniques to narrow them down. But if one or two of them describe minor, puny, tiny, relatively insignificant stuff, then you'll know what to do with them—get rid of them!

11. Opposites attract.

If a set of answer choices contains two opposites, one of them is usually the correct answer. Notice that we said "usually." In other words, use the other Process of Elimination techniques first, but in the end, if you see two opposites, pay attention.

12. Two rights make a wrong.

If a question has two very similar answer choices, many times they both are wrong for the reason that they both can't be right. Be very careful with this technique, though. Make very sure that the two answer choices are, in fact, very similar! If there's any difference between them, the correct answer to the question might hinge on that difference. If, however, the difference between them isn't at issue in the question, then eliminate them both because you can't keep them both.

13. Why do they even bother?

If all else fails and you're guessing at a correct answer, try to figure out why the test writers are even bothering to ask the question in the first place. For example, if a question asks you to identify a characteristic of a society in ancient Asia, you can probably eliminate an answer choice that says, "The society had very little impact on the development of Asian culture." Why can you eliminate it? Because the test writers aren't going to ask a question about a society that didn't have an impact. In fact, if the society hardly had any impact at all, the test writers might not even know about it! Here's another way of looking at it: The test writers developed this test to make sure that you have a firm grasp of world history and geography. Therefore, every culture, event, and individual tested *must* be significant to your world history or geography education. If you can remember to ask yourself, "Why is this question important?" instead of "What is the answer to this question?" then you'll be able to eliminate answer choices that are inconsistent with the goals of the test.

Maps, Cartoons, Charts, and Other Stuff That Takes Up a Lot of Space

You'll notice that each test will include a bunch of questions that refer you to a map, cartoon, or chart. These questions are very similar to all the other multiple-choice questions, but there are a few things in particular that you should keep in mind. First, never pick an answer choice that is inconsistent with your knowledge of world history and geography, regardless of the chart or graph or cartoon. For example, even if you don't understand the information in a chart, you shouldn't pick an answer choice that claims that world population is declining because it isn't. Even if a map confuses you, you can still eliminate answer choices that describe events that never occurred, just like you can with other multiple-choice questions.

Second, after you've eliminated answer choices that just aren't true, try to focus on the information presented on the map or on the chart. There might be several answer choices that make sense and that are consistent with your knowledge of world history and geography. You need to find the answer choice that the information on the map or in the chart describes.

> A review of charts, graphs, maps, and cartoons is included in the Subject Review portion of this book.

Finally, be especially careful with cartoons. In almost every cartoon question, one of the answer choices describes a literal translation of the cartoon. Don't be literal. Cartoons employ symbolism. Therefore, while the correct answer needs to *represent* the event in the cartoon, it will not describe the cartoon literally.

There's Just No Guessing About Guessing

The reason that this chapter includes so many Process of Elimination techniques should be self-evident: You'll need them because you're going to be guessing. You've heard it before, but it's worth repeating: *Don't leave any questions blank!*

That said, there's something else you need to keep in mind: If a particular question gives you difficulty, you might want to skip it *temporarily*. It's possible that after you answer other questions, you'll be in a better place to evaluate the question you skipped. Also, you don't want to get all huffy-puffy after just a few questions because there are about 70 more questions to answer, and most of them are probably not quite as difficult. So, if you come across a question that is particularly challenging, move on and return to it later.

Get the Points

Remember: A point is a point is a point. All questions are worth the same number of points. As such, it doesn't matter which questions you answer correctly as long as you get enough of them right. Keep this in mind as you review the subject review chapters. We're going to tell you which stuff is extremely important and which stuff isn't as important. If you're just hoping for a passing score, focus on topics that are tested frequently. If you're hoping for special recognition, you'll need to learn just about everything.

Take a Deep Breath. Relax. Become One with the Techniques.

This chapter is crammed with stuff that will help you score well on the exam. You might want to read it a couple of times. In the end, don't underestimate the power of these test-taking techniques. The real demonstration will occur when you take the practice test at the end of the book (don't forget to read the subject review chapters first!). After you take the practice exam, read the explanations for the questions that you answered incorrectly. The explanations will tell you which of the techniques from this chapter you could have used to eliminate answer choices. Once you understand why the right answers are right and the wrong answers are wrong, you will be test-wise and ready for the real test!

It might help to review the Process of Elimination techniques before the exam, to freshen your memory.

Chapter 3

Review of World History until AD 1000

First Things First. Let's See What You Know

If you need to, review the Process of Elimination techniques that were covered in chapter 2, then practice using them in the practice quizzes.

Take the following quiz just so you can see what you remember from your world history and geography classes. Go ahead, don't be shy. Guess if you have to, just like you'll do on the real test.

1   All of the following were true of homo sapiens during the Paleolithic Era EXCEPT

   A   They learned to use and control fire.

   B   They developed an oral language.

   C   They developed a written language.

   D   They domesticated animals.

2   Hieroglyphs are associated with which one of the following civilizations?

   F   ancient Egypt

   G   Mesopotamia

   H   Indus valley

   J   Shang China

3   All of the following are true of the major ancient river-valley civilizations EXCEPT

   A   They all had a surplus of food.

   B   They all traded with the regions that surrounded them.

   C   They all built cities.

   D   They all built pyramids or ziggurats.

4   The Greeks were polytheistic. Polytheism is

   F   a social structure in which most of the people are slaves

   G   a culture in which the latest technologies are always used

   H   a religion in which many gods are worshipped

   J   the belief that work is more important than the arts

5   The Peloponnesian Wars were fought

   A   between Athens and Sparta for control of the region

   B   between Athens and Carthage for control of the Mediterranean

   C   between Athens and Rome for control of trade routes

   D   between Athens and the Macedonians for control of lands to the East

6 The laws of Rome were codified and called

   F   the Ten Commandments

   G   the Code of Hammurabi

   H   the Twelve Tables

   J   the Constitution

7 As the Roman Empire declined, which of the following occurred?

   A   The Roman citizens rushed to the defense of Rome.

   B   The military became highly organized and focused its efforts on defending the western and northern frontiers.

   C   Diocletian divided the Empire into an eastern half and a western half.

   D   The influence of the Catholic Church decreased dramatically.

8 Which of the following is least associated with the Islamic faith?

   F   the Five Pillars

   G   the Byzantine Empire

   H   the Qur'an

   J   monotheism

9 During the European Middle Ages, what did the peasants get in return for working the land in the fief?

   A   money

   B   part ownership of the land

   C   protection and a place to live

   D   education and training in the cities

10 What leader reunited parts of the old Roman Empire, including parts of Germany and France, with the blessing of the Roman Catholic Church?

   F   Verdun

   G   Justinian

   H   Constantine

   J   Charlemagne

11 Under the caste system, an individual

   A   can advance up the social ladder during this life if he or she works hard

   B   is born into a particular caste based on luck

   C   must live according to the duties of his or her given caste to be rewarded with a higher caste in another life

   D   may be asked to join a higher a caste if he or she pleases people in higher castes

12 The Silk Road was built primarily to

   F   serve as an escape route for Chinese from invading tribes

   G   allow T'ang city-dwellers easy access to farmlands

   H   make it easier for Buddhists to make pilgrimages to holy Buddhist sites

   J   provide a link between China and the West for trade purposes

13 Which two religions did most people in traditional Japan follow?

   A   Buddhism and Hinduism

   B   Buddhism and Shintoism

   C   Shintoism and Hinduism

   D   Buddhism and Christianity

14 Ancient Kush and Ghana in Africa were significant to world history because

   F   They were the birthplaces of Islam.

   G   They traded with the Middle East and therefore opened otherwise-isolated parts of Africa.

   H   They developed a complex calendar system that became the model for the modern Western calendar.

   J   They are generally recognized as the first democracies.

15 Mesopotamia is in modern-day

   A   Iraq

   B   Israel

   C   Egypt

   D   India

Score Yourself

Okay, now score yourself using the answers at the end of this chapter on page 70. If you answered more than 12 correctly, you're in great shape. If you answered between 8 and 12 correctly, you need to do a little studying. If you answered fewer than 8 correctly, read this chapter *completely*! All the stuff that's most likely to show up on your test is right here in this chapter. Read it. Then read it again. If you understand it, you'll do just fine on test day.

Keep in mind that this quiz has only 15 questions, whereas the real test will have 71 questions, 61 of which will count toward your score. This quiz is merely a sample of what you are expected to know.

> Even if you did well on this quiz, you are still responsible for knowing all of the information contained on the following pages because any of it might show up on the real test.

Stuff You're Expected to Know

There are five broad categories of information from your eighth-grade history class for which you are responsible. You will also need to know geography from your 10th-grade geography class (geography will be reviewed in chapter 4). The five broad categories are:

1	Ancient civilizations	7	questions
2	Greece and Rome	11	questions
3	Middle East, Russia, and early medieval Europe	9	questions
4	Asia, Africa, and the Americas	8	questions
5	History skills	8	questions

And two more categories will be covered in chapter 4:

6	Geography knowledge and concept	12	questions
7	Geography skills	6	questions

You are expected to a) know the basic facts, individuals, groups, and concepts from each of the categories; b) understand the significance of those facts, individuals, groups, and concepts; and c) understand the "big picture." If you review everything in this chapter a few times, you'll be well prepared for the 43 questions from these five categories.

How to Study

If you have the time, read this chapter chronologically because history builds upon previous events. You'll understand how world civilizations developed if you read the information in order rather than skipping around for random facts. However, if you already have a big-picture understanding of world history and just need to fine-tune a few key points, use the headings and the words printed in bold to guide your study. For example, if you don't remember much about Charlemagne, scan through the chapter until you find him. If the Mayan civilization escapes your memory, you'll find it in the fourth section of this review.

In the end, you need to be familiar with just about everything in this chapter if you expect to do well on the exam. We've tried our best not to waste your time. Everything included in this chapter is here because the state of Virginia has indicated that these topics, people, places, and events are fair game. What's more, there's a lot of important history that isn't included in this review, for precisely the same reason: the state of Virginia has *not* indicated that it will show up on the test! Okay? Ready? Let's start reviewing.

1. Ancient Civilizations

You will have to answer seven questions about ancient civilizations. Actually, one or two of the questions may concern human life prior to the development of civilizations. There are two big categories of information that you need to review:

1. The development of mankind from the Paleolithic Era to the Agricultural Revolution

2. The development of ancient river-valley civilizations

By reviewing the material, you increase your chances of getting all seven ancient civilizations questions right!

Below is a review of information you need to know to do well on the seven ancient-civilizations questions. Remember, even if you know this part of history fairly well, it won't hurt to review the information.

The Paleolithic Era and the Agricultural Revolution

Imagine early people. *Really* early people. They hadn't built cities. They didn't know how to farm. They didn't even have a language. They just existed, satisfying their most basic needs: food and shelter.

To think of these people (sometimes the test writers refer to them as "human ancestors"), you have to think way, way, way back in time. Think of three or four million years ago, when our ancestors were first learning to walk upright. They walked around for millions of years, not accomplishing very much by today's standards. But fortunately, these human ancestors kept evolving and evolving, and finally about 400,000 years ago (and some think as recently as about 100,000 years ago), they were so distinct from the rest of the natural world that they are now considered *homo sapiens*, which is the scientific name given to humans. Today's humans are still known as *homo sapiens*.

Early humans were constantly on the move, and they had to do it all on foot (and they didn't have tennis shoes, arch supports, or even toenail clippers!). They migrated out of Africa and made their way around the globe, toward Asia and Europe, and from Asia to Australia and the Americas. How did they do it?

Well, it took a really, really long time. It's not like one person started out in Africa and started walking until he got to Alaska. Instead, it happened over hundreds and hundreds of generations, with each generation slowly pushing outward toward new opportunities in new places, or to escape bad situations in old places. These early people were **nomadic**, which means that they were always following herds of animals (which they hunted) or trying to find new sources of water, food, or shelter. They also gathered seeds and nuts to eat, so they had to move around as the seasons changed because food wasn't always available in the same places. Since some of them hunted food and some of them gathered food, we call these early humans **hunters and gatherers**.

Eventually, they figured out that their hunting and gathering would improve if they had a way of communicating with each other. For a long time, they didn't develop written communication. But some groups *did* develop oral languages, which gave each group a strong identity of its own. These groups, which traveled together, hunted and gathered together, and communicated together, were known as **clans**.

> Groups that traveled, hunted, and communicated together were known as *clans*.

During the Paleolithic Era, humans began to develop tools that helped them to hunt, cut, and dig. They also learned to use and control fire, which helped with cooking and keeping warm. Plus, it helped them to see at night! Also, they domesticated animals, which means they actually raised animals that they could use and eat, rather than following and chasing wild animals wherever the animals went. Of course, they had to make sure they led their domesticated animals to food and water, which meant they were still nomadic, but the point is that humans began to control their environment and living conditions, rather than being controlled *by* the environment.

Sometimes, the test writers will break this era down into two time periods: the **Old Stone Age** and the **New Stone Age**. Try to remember which developments occurred during each time period. In the Old Stone Age, people developed an oral language, started using animal hides for clothing, figured out how to control and use fire, and constructed simple tools and weapons. In the New Stone Age, people domesticated animals, constructed advanced tools, figured out how to make pottery out of clay, and started weaving.

The big development, though, was learning how to farm using seeds. This was so huge that we sometimes call it the Agricultural Revolution. It was a very big deal.

Why all the fuss? Well, think about it. If you know how to farm, you can stay in the same place. You don't have to move to find food and water for yourself or your animals. As long as you have a relatively constant supply of water, reasonably predictable weather, good soil, tools, and the will to work hard, you can be a farmer.

Remember to use your new POE techniques to eliminate answer choices when you're not sure which choice is correct.

Think of the consequences! A steady supply of food! All in one place! And to top it off, if you grow enough food, there might be extra for some other people. Which means that everybody doesn't have to be gathering and hunting food all the time. Some people can farm lots of food for everybody, and then others can focus on other stuff . . . like building cities.

Quick Quiz 1

1 **All of the following were true of homo sapiens during the Paleolithic Era EXCEPT**

   A They learned to use and control fire.

   B They developed an oral language.

   C They developed a written language.

   D They domesticated animals.

2 **Groups of early humans were nomadic. Nomadic means that people**

   F frequently move from place to place in search of basic needs

   G have not yet developed a written language

   H are farmers

   J know how to use and control fire

3 **Human ancestors developed into homo sapiens approximately**

   A   1 to 4 million years ago

   B   100,000 to 400,000 years ago

   C   1,000 to 4,000 years ago

   D   100 to 400 years ago

Check your answers to Quick Quiz 1 on page 70. If you missed any of these questions, reread the previous section.

Ancient River Valley Civilizations

All of the world's original great civilizations were located in river valleys. And when you think about it, it makes sense. Rivers provided a regular supply of water, which is necessary for human survival. They also tend to be surrounded by soil that is loaded with nutrients deposited when the river floods. Rivers are also a great means of transportation, especially before trains, planes, and automobiles were invented.

Civilizations were able to develop because people stayed in the same place year-round. As some of the clans began to settle in river valleys, they grew crops, which created a food surplus, which meant that there was more food than the farmers themselves could eat. This, in turn, allowed some of the people to farm while others developed governments, built homes and cities, established religions, and so on. With the advent of irrigation, greater areas could be farmed, which led to greater food surpluses, which led to the development of entire complex civilizations.

The test writers will focus on four ancient civilizations: 1) ancient Egypt, 2) Mesopotamia, 3) Indus Valley, and 4) Shang China. You may be expected to place these four civilizations on a map and know about each civilization's contributions to agriculture, economics, government, religion, language, and math/technology. You will also need to understand the social structures within each civilization. All the details you need are included below.

Read on to learn what you'll need to know to answer questions about ancient river-valley civilizations.

Ancient Egypt (3500–1069 BC): A test writer's favorite

The ancient Egyptian civilization developed along the Nile River, where the soil was rich and the agricultural opportunities were plentiful. The Nile cuts through an otherwise arid landscape, so the people of ancient Egypt clustered along the river's banks, where in addition to farms, they constructed towns and cities.

The Nile flooded annually, which was good news for the civilization because the floodwaters deposited rich layers of topsoil with which to grow crops. But flooding was also a problem since most people lived along the banks of the river. Cooperation in managing the impact of the floods, then, was essential for the civilization to thrive.

The entire river valley united under King Menes, who built his capital at Memphis and led efforts to manage the floodwaters and build drainage and irrigation systems. As a result of the unification, the Egyptian civilization became extremely wealthy and powerful. Rulers, known as pharaohs, directed the construction of the pyramids, which were enormous tombs for the pharaohs. In addition, an advanced writing system known as **hieroglyphs**, which consisted of a series of pictures that represents letters and words, allowed the Egyptians to communicate events and ideas not just to each other, but to later generations (we *still* learn about ancient Egypt by deciphering hieroglyphic messages!). The messages were usually either carved into stone or written on papyrus.

The history of ancient Egypt is actually divided into three kingdoms: Old Kingdom, Middle Kingdom, and New Kingdom. Each kingdom had its rise and fall. It's doubtful that you will be asked about the details of each kingdom, but you should be aware that over a period of nearly 2,500 years, ancient Egypt had many ups and downs and power struggles.

Over time, the civilization became dependent on trade because the people needed a constant supply of timber and stone for ambitious building projects, and because they craved luxuries such as gold and spices. Trading had an enormous impact on the region because it increased the spread of ideas throughout the known world and brought civilizations into contact with each other. It also led to enormous wealth for the ruling class of Egypt. The pharaohs became obsessed with the acquisition of worldly goods, and since their religious beliefs encompassed an afterlife in which you could take your possessions with you, the pharaohs insisted on being buried with their most prized possessions, including items as trivial as jewelry and as cumbersome as chariots.

Imagine being buried with your car. The pharaohs believed that your possessions crossed into the after life with their bodies, so they wanted to bring their best stuff along for the ride.

The ancient Egyptians also had a great appreciation for the arts and for learning. These ancient civilizations were the first places where culture could develop. Prior to the existence of civilizations, everybody needed to hunt or gather, so there was no time for the arts, advanced learning, or cultural practices and institutions.

To say the least, ancient Egypt was *the* center of activity in the region for centuries. But it would not last forever. The central government of the Egyptian New Kingdom eventually collapsed in the wake of the expansive Assyrian Empire.

Mesopotamia (4500 BC–2300 BC): Things were super in Sumer

Mesopotamia literally means "land between the rivers." The rivers were the Tigris and the Euphrates, and a series of ancient civilizations—most notably Sumer and Babylon—thrived along their banks. Mesopotamia is part of a larger area of relatively arable land known as the Fertile Crescent, which extends westward from Mesopotamia toward the Mediterranean.

Sumerians were the first to settle the region. Using a writing system known as **cuneiform**, they made tremendous contributions to the development of math and science.

Babylon was the second great civilization in Mesopotamia, named after the people who conquered the Sumerians. One of the Babylonian rulers, Hammurabi, developed the now-famous **Code of Hammurabi**, which established rules for every aspect of social life. This code of conduct, for the first time, extended some limited legal protections to all the people of the empire, except for the slaves. When Mesopatamian cultures won new territories through warfare, they enslaved the captured people, often cutting off their hair to distinguish them from non-slaves. The Babylonians also built pyramid-like structures known as **ziggurats**. Eventually, the Babylonians fell to the Hittites.

Indus Valley (4000 BC–1800 BC): Indus industry ruled

Farther east, the Indus Valley (in present-day India) was colonized by farmers in the late fourth millennium BC. Early settlements were replaced by planned towns and cities, which focused on early forms of industry. Among the many contributions of the Indus Valley civilization were pottery, flint blades, and copper products.

Still, compared with other civilizations of the time, the Indus Valley civilization remained relatively mobile, with many inhabitants moving their herds to graze. Consequently, these semi-nomadic people were in a good position to carry trade to the Middle East, where they would trade Indus-made goods with those made in the Fertile Crescent. The civilization declined around 1800 BC, probably due to rivers drying up, disease, and the Aryan invasion from the North.

> You can read more about the people and geography of the Fertile Crescent in chapter 6.

Shang China (around 1700–1100 BC): Shang on the Hwang

Shang China developed in the Hwang Ho River valley, and, like the other civilizations, used its stable agricultural surplus to build a trade-centered civilization. Its bronze-working was exceptional, but what really set Shang China apart from other groups in its region was its advanced writing system. This system created a relatively literate population, which meant knowledge expanded with each successive generation.

At its height, the civilization controlled large parts of northern China. The first Shang king, Tang, achieved dominance by defeating 11 groups of people and then winning more than 36 other groups simply by proving that he was a fair ruler and really good guy.

Shang China was also militarily powerful. Thousands of workers built walls around the towns and cities along the Hwang Ho River. Warriors used chariots to overrun enemies. Still, Shang China fell, in part due to the tyrannical excesses of the very last Shang ruler, Chou. His cruelty to his own people was so extreme that many of the Shang people were not disappointed when he was finally defeated at the hands of the Zhou around 1100 BC.

After Shang China fell, the Zhou became China's longest-ruling dynasty, controlling the region for nearly 900 years.

What all four civilizations have in common

Keep in mind that all four of these civilizations developed in river valleys and succeeded because of agricultural surpluses. They all developed food storage systems to maximize the benefit of the surpluses.

All four civilizations also grew wealthy in part because of trade. Trade allowed them to build cities and support the arts. As wealth grew, centralized governments increasingly controlled the economies. The governments even developed accounting systems to keep track of their vast empires.

> It's as important to know what the four ancient civilizations had in common as it is to know how each civilization was unique.

In addition, these four civilizations practiced **polytheism**, which means they believed in many gods as opposed to one God. They also had written codes of law, most notably the Code of Hammurabi.

Finally, all of these civilizations used metal for tools, built chariots for warfare, constructed plows for farming, and developed mathematics to aid in building.

A Quick Overview of Civilizations that Came Later

You might be asked about the Persians, Hebrews, and the Phoenicians. The Persians dominated present-day Iran around 500 BC. For the purposes of the test, they're important because they fought the Greeks for control of the eastern Mediterranean (which will be discussed later). The Hebrews settled in Palestine. Unlike every other civilization of the time, Hebrews, a Jewish people, were monotheistic, not polytheistic. The Phoenicians settled along the coast of the Mediterranean in a collection of city-states. The most important accomplishment of the Phoenicians was the development of the alphabet on which Greek, and later many modern languages, would be based.

> You think you understand enough about ancient civilizations? Take the Quick Quiz to make sure.

Carthage was a wealthy Phoenician trading center and gradually established its own empire. As it expanded, however, it clashed with two powerful empires that were coming into their own on the northern banks of the Mediterranean: the Greeks and the Romans.

Quick Quiz 2

1 **Hieroglyphs are associated with which of the following civilizations?**

   A ancient Egypt

   B Mesopotamia

   C Indus Valley

   D Shang China

2 **The first alphabet was developed by**

   F the Egyptians

   G the Babylonians

   H the Shang Chinese

   J the Phoenicians

3 **Cuneiform was a**

   A burial procedure developed by the Egyptians

   B writing system developed by the Sumerians

   C architectural style developed by the Babylonians

   D religious practice developed by the Hebrews

4 **Each of the following is true of major ancient river-valley civilizations EXCEPT**

   F They all had a surplus of food.

   G They all traded with the regions that surrounded them.

   H They all built cities.

   J They all built pyramids or ziggurats.

> Check your answers to Quick Quiz 2 on page 70. If you missed any of these questions, reread the previous section.

2. Greece and Rome

Some people claim that to really understand the Western world, you have to understand ancient Greece and Rome. Certainly, the test writers would agree: They'll devote 11 exam questions to the topic.

What's the big deal about Greece and Rome? Well, simply put, Western civilization essentially began in these two places. The Babylonians, the Egyptians, the Hebrews, and the Phoenicians set the wheels in motion, but the Greeks and Romans brought it all together and gave civilization a decidedly Western slant. Perhaps their most important contribution is the Western concept of representative government, but the Greeks and Romans also made important contributions to art, architecture, literature, science, and philosophy.

Because Greece and Rome are equally important, people tend to get them confused. Make sure this doesn't happen to you!

We don't know how the test writers will split the questions between Greece and Rome. On some tests there might be five questions on Greece and six on Rome; on another, it might be reversed. Make sure you don't get the accomplishments of Greece mixed up with the accomplishments of Rome. A question about Rome may have an answer choice about Greece, and vice versa.

Greece

Geography: Harbor no doubts because we have plenty

Ancient Greece was located on a peninsula between the waters of the Aegean and Mediterranean seas. There wasn't much of a possibility of growing crops, as was done in the ancient river-valley civilizations, since the land in Greece was mostly mountainous. But Greece had natural harbors and mild weather.

So how did the geography of Greece impact the development of such a dominant civilization? Well, for one thing, its coastal position aided trade and cultural diffusion by boat, which is precisely how Greece conducted most of its commercial activity. The Greeks could easily sail to Palestine, Egypt, and Carthage. The Greeks exchanged wine and olive products for grain. Eventually, the Greeks became so able at trading and commerce that they replaced the barter system, by which they traded one type of goods for another, with a money system. As the center of all this commercial activity, Athens became extremely wealthy.

Greece's limited geographical area also contributed to its dominance. Since land was tight, Greece was always looking to establish colonies abroad to ease overcrowding and gain raw materials. This meant that the Greeks had to be militarily powerful. It also meant that they had to establish sophisticated methods of communication, transportation, and governance.

Ancient Greece was militarily powerful, well situated geographically, and highly social.

Finally, Greece's mild climate promoted lots of outdoor activity, which brought people together regularly. A high degree of social contact led to the development of a very conscientious civil life, which the Greeks incorporated into their political and social philosophies.

Social structure and citizenship: It took a polis . . .

Greece wasn't a country in the way we think of countries today. Instead, it was a collection of city-states. Each city-state was known as a **polis**, which shared a common culture and identity. Although they all were part of Greece, they were independent from, and often in conflict with, each other.

The two main city-states were **Athens** and **Sparta**. Athens was the political, commercial, and cultural center of the empire. Sparta was an agriculture and military region. Most citizens in Sparta, who were known as *helots*, lived an almost slave-like existence. All the boys received heavy doses of military training, which stressed equality without individuality.

Each city-state, or polis, was composed of three groups: 1) citizens (adult males, often engaged in business and commerce), 2) free people with no political rights (mostly women), and 3) noncitizens (slaves, who accounted for nearly one-third of the people in Athens, and who had absolutely no rights). Among the citizens, civic decisions were made openly in debates. All citizens were expected to participate. In this way, Athens is seen as the first democracy, and to the extent that males could participate, it was a democracy. Since the opinions of the bulk of the adult population were not heard, however, it did not resemble a modern democracy.

Democracy in Athens did not develop immediately. Instead, as Athens grew more and more powerful, the government in Athens changed from a monarchy to an aristocracy, and finally to a democracy. The test writers might ask about two aristocrats named **Draco** and **Solon** who worked to create a democracy in Athens. Both of these men worked within the legal system in Athens to create fair, equal, and open participation among citizens.

Greek mythology: Many gods

The Greeks were polytheistic, which means that they believed in many gods. The stories surrounding these gods, like the myths of Zeus and Aphrodite, are richly detailed and are still told to this day. Greek mythology remains part of our Western heritage and language. Every time we refer to a task as *Herculean* or read our horoscopes according to our zodiac sign, we're using our Greek cultural roots.

War with Persia: Greeks hold on

Prior to the development of the democracy in Athens, Greece was involved in a series of wars that threatened its existence. The Persian Wars united all the Greek city-states against their mutual enemy, Persia. Much of Athens was destroyed during these wars, but Greece held on and the wars ended in a stalemate. Two huge victories by the Greeks, one at Marathon and the other Salamis, allowed the Greeks to maintain control of the Aegean Sea. The fact that Athens held fast against the mighty Persians has had a huge impact on Western development. Had the Persians successfully defeated Greece, many of the cultural, political, and philosophical accomplishments of the Greeks may have never occurred. Western civilization would have developed, of course, but with a much more Persian flavor.

Golden ages are always periods of economic, intellectual, and cultural enhancement. There is no such thing as a golden age when famine and strife occurred.

With Persia held back, however, Greece was free to enter into an era of peace and prosperity, which is often called the golden age of Pericles.

The Golden Age of Pericles: Athens wows the world

Under the leadership of Pericles, Athens became a cultural powerhouse. It was under Pericles that democracy for all adult male citizens was firmly established. It was under Pericles that Athens was rebuilt after its destruction by the Persians— the Parthenon was built during this reconstruction. It was under Pericles that Athens established the **Delian League** with the other city-states of Greece as an alliance against aggression from its common enemies. And, it was under Pericles that philosophy and the arts flourished. And they continued to flourish for the next two centuries.

> Golden ages are always periods of economic, intellectual, and cultural enhancement. There is no such thing as a golden age when famine and strife occurred.

Perhaps the most famous of all Greeks are the philosophers: Socrates, Plato, and Aristotle. Although our modern understanding of the world differs in many ways from theirs, these guys are still revered today as brilliant minds and the fathers of rational thinking.

During the golden age, Greek drama was dominated by the comedies and tragedies of Aeschylus and Euripedes; the sculptures of Phidias adorned the streets; and Greek architecture earned its place in history with its distinctive Doric, Ionic, and Corinthian columns. Math and science thrived under the capable minds of Archimedes, Hippocrates, Euclid, and Pythagoras (if you've taken geometry, you probably remember the Pythagorean theorem. Guess where *that* came from).

It's not that cultural achievement was nonexistent prior to the golden age. Homer, for example, wrote the epic poems *The Illiad* and *The Odyssey* before the rise of Pericles, and these two poems are widely regarded as Western civilization's first two master works of literature. But make no mistake about it: During the golden age, the arts and sciences became firmly cemented into the Western consciousness. The accomplishments of this period served as inspiration for the European Renaissance and Enlightenment nearly two millennia later and still resonate today. Perhaps that's why the test writers will be asking you questions about the golden age!

Trouble ahead for Athens

After Athens became the Greek equivalent of a super-power and dominated the other city-states in the Delian League, Sparta became resentful. Increasingly, the two city-states fought against each other rather than as allies against a common enemy. Pericles pushed Athens into a catastrophic war with Sparta, and we call this the **Peloponnesian War**.

The Peloponnesian War was a disaster for Athens because Sparta won. Yet, Sparta didn't ultimately dominate the region. Instead, both Athens and Sparta were so weakened by the war that they became vulnerable to outside aggression. The Macedonians took advantage and invaded Athens. **Philip of Macedon** conquered the region, but rather than destroying Greek culture, he respected it and allowed it to flourish.

The Macedonians didn't stop with Greece. Philip's son, **Alexander the Great**, who was taught by Aristotle, continued the Macedonian dominance. Under Alexander, the Macedonians conquered the mighty Persian Empire and then pushed eastward, creating the largest empire of the time, stretching all the way into modern-day India.

The amazing thing about the Macedonian Empire is that it essentially adopted Greek customs and then spread them to much of the known world. As such, much of the world was connected under a common law and common trade practices. *Hellenism* (the Greek world), then, didn't die with Athens, but extended beyond its original borders with the help of the Macedonians.

When Alexander the Great died at the age of 33, his empire started to crumble. With the Macedonians focused on the East, the door was open in the West for a new power to rise to the world stage: the Romans.

> Civilizations rarely come to an end of their own volition. Usually, one civilization forces the demise of another and rises to takes its place.

Quick Quiz 3

1 **Greek city-states had three groups of inhabitants. Which group was comprised of adult males who engaged in business and commerce?**

   A citizens

   B free people

   C noncitizens

   D Spartans

2 **The Greeks were polytheistic. Polytheism is**

   F a social structure in which most of the people are slaves

   G a culture in which the latest technologies are used

   H a religious system in which many gods are worshipped

   J the belief that work is more important than the arts

3 Which of the following did NOT occur during the golden age of Pericles?

    A Democracy for all adult male citizens was firmly established.

    B The Parthenon was built.

    C Athens won the Peloponnesian War.

    D The arts and sciences thrived.

4 The Peloponnesian Wars were fought

    F between Athens and Sparta for control of the region

    G between Athens and Carthage for control of the Mediterranean

    H between Athens and Rome for control of trade routes

    J between Athens and the Macedonians for control of lands to the East

Check your answers to Quick Quiz 3 on page 70. If you missed any of these questions, reread the previous section.

Rome

Geography: Safe and snug

Rome was a safe place, all things considered. The Alps to the North, helped to protect Rome from an invasion by land. And the sea surrounded the Italian Peninsula, limiting the possibility of an attack unless an entire armada floated across the sea. Yet, although Rome was isolated, it was also at a crossroads. Rome had easy access to northern Africa, Palestine, Greece, and the Iberian Peninsula (modern-day Spain and Portugal). So it was easy for Rome to reach out to the world, and just as easy for it to retreat into the protection of the mountains and sea.

The impact of geography should be clear. Since Rome was protected, it could develop into a great civilization without too much interference. Since it was well positioned with access to other areas, it could spread its influence into much of the known world.

Roman mythology: More gods

Like the Greeks, the Romans were polytheistic. Their mythology was full of gods like Apollo and Venus. Every time you see a Cupid on a Valentine's Day card, you see the impact of Roman mythology on our world today. Cupid, of course, was the Roman god of love. The Romans, as you may have guessed, didn't invent polytheism by themselves. Many Roman gods and goddesses were based on Greek mythology.

Social structure in Rome: Organized to the hilt

The class structure of Rome consisted of patricians (land-owning nobles), plebeians (all other freemen), and slaves. Does this sound familiar? It should. It's very similar to the social structure of ancient Greece. Roman government was organized as a **representative republic**. The main governing body was made up of two distinct groups: the Senate (which was comprised of patrician families) and the Assembly (which initially was made up of patricians, but later opened to plebeians). Two consuls were annually elected by the Assembly. The consuls had veto power over decisions made by the Assembly. This structure was much more stable than the direct democracies of the Greek polis, in which every male citizen was expected to participate on a regular basis. In a republic, the people had representatives, so they didn't have to vote on every issue.

This is similar to the constitutional democracy we have in the United States. Since everyone votes for representatives, it's correct to call our system a democracy. But our representatives in Congress vote on all the major issues, so our system of government is very much a republic. Indeed, the structure of our government was modeled on the system used in the Roman Republic. (Instead of two consuls, though, we have one, known as the president.) Early on, Rome developed civil laws to protect individual rights (in some ways similar to our Constitution). The laws of Rome were codified and became known as the **Twelve Tables of Rome** (the concept of "innocent until proven guilty" originated here). Later, these laws were extended to an international code that Rome applied to its conquered territories.

> *Remember:* In a direct democracy, every citizen participates in government first-hand. In a republic, citizens elect representatives to participate in government for them.

Roman military domination: All directions, all the time

As Rome expanded, Carthage, a city in North Africa with ambitions of its own, became its first enemy. The two cities couldn't stand each other, so it didn't take long for their hatred to escalate into full-fledged war. The wars between Carthage and Rome were called the **Punic Wars**. The First Punic War was for the control of the island of Sicily, and Rome won. The Second Punic War began with a Carthaginian attack by **Hannibal**. In an amazing feat, Hannibal took his army all the way to northern Italy, crossed the Alps (with elephants no less!) and surprised the Romans, who were expecting them to attack from the South. Hannibal invaded Rome, but he wasn't able to destroy the Roman military. Eventually, Rome defeated Carthage and forced it to surrender its navy and all of its foreign holdings. The Third Punic War was instigated by Rome 50 years later. Rome burned Carthage to the ground. With Carthage out of the picture, Rome had a much easier time expanding throughout the Mediterranean. Rome went on to obtain Greece by conquering the Macedonians. They also fought the Gauls to the north and the Spaniards to the west. All of this conquest spread Roman culture (which itself was linked to Greek culture) throughout much of western Europe, the Middle East, and northern Africa. To maintain their vast empire, the Romans built roads, aqueducts, and greatly expanded their navy. As a result, Rome became very, very wealthy due to trade and conquest.

The collapse of the republic and the rise of imperialism

Following the Punic Wars, while Roman influence was expanding, the situation in and around Rome was becoming unsettled. Several events caused this restlessness. First, large landowners started using more slaves from the conquered territories. This pushed small farmers out of business and into cities. The cities were suddenly crowded with too many plebeians and not enough jobs. A high rate of unemployment resulted. Second, the Roman currency was devalued, so there was a high rate of inflation. This meant that the plebeians did not have enough money to buy the things they previously could. Third, political leaders were fighting each other. The power of the Senate weakened and was transferred to three guys known as the **first triumvirate**: Pompey, Crassus, and **Julius Caesar**.

Caesar was given power over southern Gaul (France) and other parts of Europe, but he chose not to conquer Germany (which was important because Germany developed a different culture and would later serve as a training ground for groups intent on conquering Rome). Civil war between the Senate and Caesar pushed Pompey and Crassus out of the picture, and Julius Caesar became emperor for life. His life didn't last long, though. The angry senators assassinated him.

A *triumverate* is a rulership by three people (kind of like a *tri*angle has three sides).

After the death of Julius Caesar, a **second triumvirate** of Octavius, Marc Antony, and Lepidus came to power. This whole triumvirate idea was obviously a bad one, and it didn't improve the second time. Power again shifted to one person. Octavius rose to power, assumed the name **Augustus Caesar**, and became emperor. Augustus was the first true emperor of Rome (Julius really didn't count) and the days of the Roman Republic were now over once and for all. Rome was now an empire led by a single emperor. The Senate and the Assembly ceased to exist.

Pax Romana: Peace and prosperity

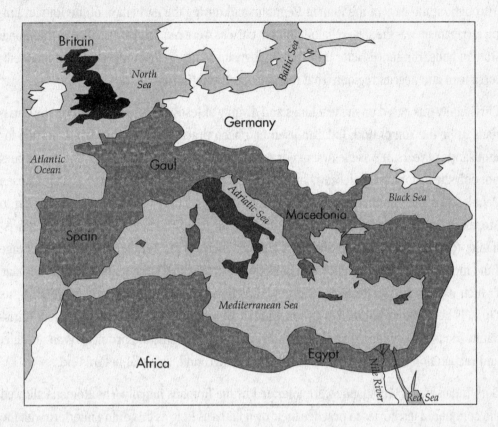

Under Augustus, Rome became the capital of the Western world. Augustus established rule of law, a common coinage, civil service, and secure travel for merchants. With everything in place throughout the empire, stability returned, and for 200 years, the empire enjoyed a period of peace and prosperity known as the ***Pax Romana*** (Roman Peace). The fascinating thing about the *Pax Romana* was that while laws, civil service, taxes, and coinage were uniform throughout the empire, many of the traditional customs of the people in the conquered territories continued. In other words, the conquered territories were handled relatively fairly, and conquered peoples were often granted the rights of citizenship. Under imperial power, the Roman Empire reached its largest geographical proportions through additional military conquests.

But the real story was in the development of the arts. For quite some time, Greece was the artsy sister of the Western world. With the Roman Peace, however, the arts in Rome flourished, especially literature (including Ovid's *Metamorphoses* and Virgil's *Aeneid*) and architecture (including the Pantheon, Colosseum, and Forum). Science, which was always a Roman endeavor, also reached new heights. Roman roads and aqueducts were feats of engineering. Meanwhile, Ptolemy looked to the heavens and greatly influenced developments in astronomy. All of these contributions during the *Pax Romana* influenced the entire Western world. Today we owe many of our artistic and scientific origins to the Romans (and Greeks).

Religious diversity: New chiefs of beliefs

Throughout the days of the Roman Republic and during the early days of the Roman Empire, paganism was the state religion. Roman citizens were required to sacrifice to traditional Roman gods. But shortly after the reign of Augustus, a new religion developed in the Mediterranean and Aegean regions. That religion was **Christianity**.

Christianity was based on the teachings and divinity of Jesus Christ, a Jew believed by Christians to be the Son of God. Judaism itself had been practiced by Hebrews in Palestine for thousands of years. It was the first major monotheistic religion. Its text, the Torah, outlines not only the history of the Hebrews but also God's covenant with the Hebrews as the Chosen People. The Ten Commandments, believed by Jews and Christians to have been given to Moses by God, outlined the moral obligations of all Jews. The new upstart religion, Christianity, took Judaism one step further. It adopted much of the Torah as a common heritage (and incorporated much of the Torah in its own text, the Bible). But it added to this heritage (which it calls the Old Testament) a new direction (the New Testament), in which Jesus Christ is accepted as humanity's savior, without whom humanity would be doomed. Christianity focused attention on redemption through faith in Jesus, the promise of an afterlife, and recognition of all believers, regardless of background, as equal before God.

Both Judaism and Christianity were tolerated by the Romans initially. The Romans allowed the conquered territories to practice their own faiths as long as this didn't interfere with the goings-on of the empire. Jewish resistance to Roman control, however, led to the suppression of Judaism. Many Jews migrated to other parts of the empire to escape persecution. And as the apostles of Jesus and missionaries tried to spread Christianity throughout the empire, the Romans increasingly saw the new religion as a threat to both paganism and their own power. **Emperor Nero** persecuted Christians to the point of killing them in open spectacles at the Roman Colosseum. But Christianity kept growing. Eventually, **Emperor Constantine** converted to Christianity and ended the persecution with the **Edict of Milan**. Christianity became the official religion of the Roman Empire and has been one of the world's most influential religions ever since.

> Followers of Judaism continued to look for a savior, while followers of Christianity believed that a savior had already come.

Christianity becomes highly organized after Nicea

Technically, Christianity began at the time of Christ. Jesus's first 12 disciples were the first Christians. But Christianity as an organized religion didn't really start until the **Council of Nicea** some 300 years later. The council's main job was to formulate what Christians now call the Bible; in other words, the council decided which ancient Jewish scriptures and which recently written texts were to be included. With the Council of Nicea, Christianity was transformed from a faith into an organized religion. With the blessing of Emperor Constantine, the religion became *the* Church. The early Christian Church organized itself in Rome and appointed a pope to be in charge, plus bishops to head districts throughout the empire, and individual priests to serve each parish or congregation within the empire. Even as the Roman Empire declined, this highly organized church structure gave the citizenry a stable environment and a strong identity.

As the Roman Empire continued it's downward spiral, the people started giving their first loyalties to the Church rather than to the empire, causing the empire to decline even further.

The decline of the Roman Empire

Slowly but surely, the Roman Empire fizzled. There were a lot of reasons for the decline. The leadership was poor. The economy had stalled. The citizenry became apathetic. The politicians were constantly fighting. It was just plain messy. Perhaps the biggest reason for the decline was that barbarians from the North kept attacking the empire, one wave after another. The cost of defending the empire meant that the leaders couldn't spend money on improvements for their own people. What's more, the military was spread too thin, so it wasn't able to defend the empire adequately. The empire was simply too huge to defend successfully.

Faced with the possibility of losing everything, **Diocletian** divided the empire into an eastern half and a western half. Since most of the invasions were occurring in the western half, the decision was made to let the invasions continue unchecked and to focus all the efforts on defending the eastern half. Constantine built a new capital (a "new Rome") at Byzantium and renamed it Constantinople. This city—modern day Istanbul—would serve as the headquarters for the **Byzantine Empire** (also known as the eastern Roman Empire). In the meantime, Rome collapsed. It ceased to have a Roman emperor in AD 476.

Quick Quiz 4

1 The Roman Senate and Roman Assembly are associated with

A   the Roman Republic

B   the Roman Empire

C   the Roman city-state

D   the Roman-Greco Alliance

2 The laws of Rome were codified and called

F   the Ten Commandments

G   the Code of Hammurabi

H   the Twelve Tables

J   the Constitution

3 **The Punic Wars were fought between**

   A the Senate and the Assembly

   B Rome and Greece

   C Rome and the Gauls

   D Rome and Carthage

4 **After the fall of the Roman Republic, who led the Roman Empire into a Pax Romana?**

   F Julius Caesar

   G Octavius

   H Marc Antony

   J Augustus Caesar

5 **Which of the following was NOT an accomplishment of Rome during the Pax Romana?**

   A Taxes and coinage were made uniform throughout the empire.

   B Literature and architecture flourished.

   C People in conquered territories were given Roman citizenship.

   D Diocletian divided the empire into an eastern half and a western half.

6 **Which religion began during the Roman Empire and was eventually adopted by Emperor Constantine?**

   F Christianity

   G Judaism

   H Islam

   J Hinduism

7 **As the Roman Empire declined, which of the following occurred?**

   A The Roman citizens rushed to the defense of Rome.

   B The military became highly organized and focused on defending the western and northern frontiers.

   C Diocletian divided the empire into an eastern half and a western half.

   D The influence of the Catholic Church decreased dramatically.

Check your answers to Quick Quiz 4 on page 70. If you missed any of these questions, reread the previous section.

3. The Middle East, Russia, and Early Medieval Europe

This section concerns religious history in the Middle East and Europe, and for good reason: The exam will include about nine questions about the Middle East, Russia, and early medieval Europe. Religion was the biggest force in the world during the early Middle Ages. Make sure you understand the differences between Judaism, Christianity, and Islam, as well as the impact of each on the world through AD 1000. Also, make sure you understand the structure and impact of feudalism on medieval Europe since you'll have at least a couple of questions on the topic.

Conflict Between the Muslim and Christian Worlds

In the seventh century, a new faith took hold in the Middle East. This faith—called **Islam**—was monotheistic, just like Judaism and Christianity. Muslims believe that Allah (God) transmitted his words to the faithful (in the form of the Qur'an) through his prophet, Muhammad. Muslims believe that salvation is won through submission to the will of God, and that this can be accomplished by means of the **Five Pillars of Islam**. These five pillars include: 1) confession of faith, 2) prayer five times a day, 3) charity to the needy, 4) fasting during the month-long holiday Ramadan, and 5) pilgrimage to Mecca at least once during one's life.

Islam shares a common history with Judaism and Christianity. It accepts Abraham, Moses, and Jesus as prophets (although it does not accept Jesus as the Son of God), but holds that Muhammad was the last great prophet. Like Christians, Muslims believe that all people are equal before God and that all should be converted to the faith. Early on, Islam split into two groups: Shi'a and Sunni. The split occurred over a disagreement about who should succeed Muhammad as the leader of the faith.

Islam's strict codes of belief are listed and described in detail in the **Qur'an** (which is also spelled *Koran*). This sacred text greatly influenced both the religious and political life of Muslim culture. The tenants of Islam became as officially practiced in Muslim culture as the tenants of Christianity were practiced in the Roman and Byzantine Empires. Islam spread rapidly through the Middle East and northern Africa, which worried Christian leaders to the North. Later, when Muslim culture spread from Africa to the Iberian Peninsula of Europe (present-day Spain and Portugal), the Christians weren't just worried, they were livid.

> The Five Pillars of Islam are confession, prayer, charity, fasting, and pilgrimage.

The Islamic Empire: Surprise from the Southeast

In the seventh century, a new empire arose out of Arabia and became one of history's great empires. The Arabs, who before converting to Islam had been relatively disorganized, surprised the world by defeating the Persian Empire and parts of the Byzantine Empire, and then expanding all the way through North Africa into Spain. The empire spread Islam through a huge swath of the Mediterranean and points east and south, and Muslim culture suddenly became all the rage. By AD 732, the Muslims reached all the way across the Mediterranean into Spain and advanced to within about 100 miles of Paris, where they were stopped by Charles Martel and his Frankish warriors at the **Battle of Tours.**

Like the Roman Empire, the Islamic Empire was often tolerant of the local customs of the areas it conquered, although Christians and Jews were often persecuted in the Levant (present-day Israel, Jordan, Syria, and Lebanon). At the same time, Arabs spread their own architecture, philosophy, and poetry throughout the empire. They built tremendous capitals at Damascus and then later at Baghdad, which became one of the great cultural centers in the world. The golden age of Muslim culture occurred during the **Abbasid Dynasty**, a time when Islamic arts reached new heights. Like all golden ages, this one produced the greatest examples of architecture, literature, and mathematical scholarship of its time, perhaps best represented by the tremendous mosque in Cordoba, Spain.

Most significant for the purposes of the test, however, are not so much the innovations of the Arabs but rather their role in preserving Western culture. The Muslims encountered the classic writings of ancient Athens and Rome, including those of Plato and Aristotle, and translated them into Arabic. While European civilizations were highly decentralized and dismissive of their ancient past, the Arabs kept the Western influences alive. Later, after the Muslims and Christians battled each other for control of the eastern Mediterranean during the **European Crusades**, which eventually led to trade and exchange, Europe "redisovered" this history.

The Byzantine Empire and Russia from AD 300–1000

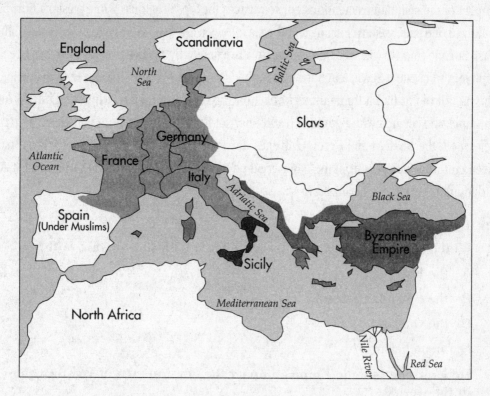

As discussed previously, Constantine the Great established the eastern capital of the Roman Empire in Constantinople, and turned the city into one of the largest, wealthiest, and most beautiful cities of its day. Constantinople became the center of religious, social, cultural, and commercial life for the Christian world, although Rome (which by this time was politically bankrupt) still tried to maintain a footing in religious affairs.

The eastern Roman Empire is known now as the Byzantine Empire. Within it, the Christian Church was preserved while Rome was falling. Under Justinian, the former glory and unity of the Roman Empire was somewhat restored. The region flourished in trade and support for the arts. Constantinople and Baghdad rivaled each other for cultural supremacy. The Justinian Empire is perhaps most remembered for the **Justinian Code**, a codification of Roman law, and its religious art work, especially in the form of mosaics, which are pieces of stone or glass combined to make a picture.

It should be noted, though, that the Constantinople and Byzantine Empires were not a simple continuation of Roman ideals. While the Byzantine Empire followed traditional Roman laws and organization, it was Greek in language and culture. Eventually, the western half of the Church reasserted itself and became increasingly independent of Constantinople. As the Middle Ages progressed, the western Church eventually split from the eastern half. In the West, the Holy Roman Empire was established under Roman Catholicism. In the East, the Byzantine Empire nourished what is known as the Greek Orthodox Church. This split still remains today.

The eastern half of the empire included parts of eastern Europe and Russia, and so the impact of the split had tremendous consequences for these regions. While western Europe followed one path, eastern Europe and Russia followed another. The Slavic peoples of southeast Europe and Russia were converted to Christianity by St. Cyril, who used the Greek alphabet to create a Slavic language that to this day is used in parts of eastern Europe and Russia. All of this had a tremendous impact on Russia. The Russian Orthodox Church, for example, was aligned with Byzantine traditions, not Roman. Later, when the Roman church reformed, the Russian and Greek churches did not. As a result, Russia is and always has been culturally different from the other great powers of Europe. To this day, Russia has had a difficult time with Westernization.

Quick Quiz 5

1 Which of the following is least associated with the Islamic faith?

   A the Five Pillars

   B the Byzantine Empire

   C the Qur'an

   D monotheism

2 How did the Islamic Empire impact the continuance of Western thought in the world?

   F The empire promoted Christianity throughout its newly acquired territories.

   G The empire preserved ancient Greek and Roman texts.

   H The empire mimicked Roman architecture.

   J The empire strictly followed the Twelve Tables of Rome.

3 Which of the following became the capital of the Islamic Empire during its golden age?

   A Damascus

   B Baghdad

   C Jerusalem

   D Mecca

4 The two main divisions in the Christian church during the Middle Ages were

   F Catholicism and Islam

   G Roman Catholicism and Greek/Russian Orthodoxy

   H Roman Catholicism and Protestantism

   J Greek Orthodoxy and Russian Orthodoxy

Check your answers to Quick Quiz 5 on page 70. If you missed any of these questions, reread the previous section.

Europe during the Middle Ages from AD 500–1000: Feudalism

Feudalism is the name of the European social, economic, and political system during the Middle Ages. The economy was mostly based on farming. The people who owned *the* land were the wealthy and powerful ones, while the people who work*ed the l*and were neither wealthy nor powerful. In fact, they lived a very meager existence.

Here's how feudalism worked: At the top, there was the king. He had power over an entire territory of land, called his kingdom. Beneath him were the nobles, called **vassals**, who, in exchange for military service to the king, had power over sections of the kingdom. So, if two kings in neighboring kingdoms got into a fight, the two kings would call up their nobles and go to war. Below the vassals were peasants and **serfs**, and there were a whole lot of them. These were the common people who worked the land but didn't own it. The serfs were given a place to live and protection from invasions, and in exchange, they worked the land to meet the needs of their lords. So, feudalism was essentially a system of corresponding rights and responsibilities between the lower classes and the upper classes, but the upper classes got the better end of the deal.

It's important to realize that countries or government systems did not exist in the way that we understand those things today. Instead, the land was up-for-grabs, and kings claimed power over certain territories. The people in one kingdom had characteristics in common, and so they were often intensely loyal just like in a modern day nation or country. But people who lived in a kingdom did not really see themselves as fellow citizens. Instead, they identified with each other based on their social and economic class.

Peasants were directly responsible to their lords, who were in charge of whole farms and farming villages, known as manors or **fiefs**. The peasant worked the land on behalf of the lord, and the lord gave the peasant a place to live and protection in case of invasion. Many of the manors were remarkably self-sufficient. Everything that was needed was produced right there. Food was harvested, clothing and shoes were made, and so on. The lord, of course, was responsible to the king, but only when the king called upon the lord to provide a service. At all other times, the lord was the boss of his own manor. When a lord died, his land usually was inherited based on **primogeniture**, which means the first-born son of the lord became the new lord. You would think, of course, that if a lord was killed, other surrounding lords might fight each other for control of the newly available manor, but that's not usually how it worked. Most of the lords (and knights, who *w*ere warriors that were also considered part of the nobility) followed the **code of chivalry**, which was basically an honor system which strongly condemned betrayal.

> If you forget what *primogeniture* means, remember that *primo* means best or first.

Charlemagne: The empire strikes back

After the Roman Empire broke up and the feudal system emerged, Europe was highly decentralized, meaning that there wasn't a strong central government or clear national boundaries. Instead, local kings and lords governed territorial units, as discussed above. Within these territories, people adhered to local laws and customs rather than to rules of an entire kingdom.

Charlemagne was a Germanic leader who tried to change all of this. Germany and France as countries didn't exist yet. Instead, people were called German, for example, because they spoke the German language and followed German cultural traditions, not because they lived in a country called Germany. Charlemagne tried to unite all the people who lived under the old Roman Empire and reestablish the empire in an organized way. The pope of the Catholic Church, which was highly involved in politics, crowned him emperor, and eventually this Empire, though it was relatively small, would be called the Holy Roman Empire. This set a precedent by establishing a relationship between the Church and the government that would continue throughout the Middle Ages.

Like the Romans, Charlemagne promoted the arts and learning; most of which took place in the monasteries under the direction of the Church. Charlemagne was very smart and powerful, but he didn't have absolute power. Instead, feudalism was still the main method by which society was structured. Even though Charlemagne had control over the entire empire, the local lords kept their power in the local territories, which were only required to answer to Charlemagne as needed. Most importantly, Charlemagne did not levy taxes. Without money, Charlemagne could not build a strong and united empire. After Charlemagne's death and the death of his son Louis, the empire divided among his three grandsons according to the **Treaty of Verdun.** Now split, the empire began to weaken, and eventually the division became France and Germany.

Invasions: A tribe for every occasion

Once Charlemagne's empire split, the whole region was vulnerable to attacks. The coastal regions, including present-day England, Ireland, France, Spain, and Russia, were invaded by **Vikings** from Scandinavia (also known as Norsemen). These conquerors were fierce warriors and seamen; they pillaged villages and often slaughtered villagers. The Vikings eventually settled these lands and became assimilated and Christianized. Meanwhile, as you read earlier, Muslims attacked Europe from the East and South. Even before Charlemagne's death, they gained a foothold in Spain. Still, the most central part of the European continent managed to protect itself. An alliance between Germany and Italy developed into a dual leadership—the empire (the emperor) on the one hand and the papacy (the pope) on the other.

Later, **Otto the Great**, using Charlemagne's reign as a model, reunited the central region and crowned himself emperor of the Holy Roman Empire. He expanded the boundaries of the empire, but it was still a collection of loosely governed territories under the feudal system. Like Charlemagne, Otto supported the arts and scholarship. Though he worked with the Church, he kept religious leaders weak. After Otto, the papacy and the empire engaged in a constant struggle for power that would last for centuries.

Quick Quiz 6

1 **In feudal Europe, who was just beneath the king?**

   A vassals

   B peasants

   C serfs

   D citizens

2 **What did peasants receive in return for working the land in the fief?**

   F money

   G part ownership of the land

   H protection and a place to live

   J education and training in the cities

3 **What leader reunited parts of the old Roman Empire, including parts of Germany and France, with the blessing of the Roman Catholic Church?**

   A Verdun

   B Justinian

   C Constantine

   D Charlemagne

4 **Which of the following was an impact of the invasions by Vikings into central and western Europe?**

   F Vikings controlled parts of southern Spain where they spread Islamic culture.

   G The Holy Roman Empire was kept relatively weak and small.

   H Christianity ended in northern and western Europe.

   J Italy and Germany became enemies.

Check your answers to Quick Quiz 6 on page 70. If you missed any of these questions, reread the previous section.

4. Asia, Africa, and the Americas

The test writers really, really, really like ancient civilizations. They also like the European and Middle Eastern Middle Ages. As for Asia, Africa and the Americas . . . well, as you can see, over the time span of about 1,000 years there are only eight questions about all three of these continents. This is probably because the developments around the Mediterranean had the largest impact on Western civilization.

But don't think that the developments in Asia, Africa, and the Americas weren't just as amazing. They were. On all three continents, empires built great cities, the arts flourished, and improvements in technology brought an ever-increasing array of products and methods to the populations. Of course, all four continents also experienced the bad stuff, especially wars, internal strife, slavery, and excesses of the ruling classes at the expense of the common person.

The test writers aren't clear how they will divide the questions among these continents, so it's best to know about the developments of civilizations on all four continents. Make sure you focus on the social and religious developments in each civilization. Here's a quick breakdown of the civilizations you need to review:

- India

- T'ang China

- Japan

- Kush and Ghana in Africa

- Mayan and Aztec civilizations in the Americas

India's Geography: All about the Khyber Pass

Classical India was well protected because of the Himalayan Mountains to the north of the Indian subcontinent, which helped protect India from invasion by land. Also, the deserts of southwest Asia made travel to and from India difficult. Still, some invasions managed to get through. For a short time, Alexander the Great's army took control of the northern part of the Indian subcontinent. More significant, however, was the Aryan invasion and the invasion of the Moslem Turks. All of these invasions came through Khyber Pass, a break in the mountains to the northwest of the Indus River valley.

Hinduism: Here today, here tomorrow

You may have already read about the ancient Indus Valley civilization that developed in India in the fourth millennium BC. If you haven't, and if you don't remember anything about it, you need to reread the section about ancient civilizations. The early Indian civilizations were invaded around 1500 bc by Indo-Aryan people from Iran. By 500 BC, a patchwork of kingdoms existed in the Ganges River valley. The customs of these Indo-Aryan kingdoms blended with the customs of the native population.

Religion and social customs have always been highly intertwined in India. Hinduism, an ancient polytheistic religion that holds that people are reincarnated (reborn into a new earthly life) after they die, was already in existence by the time the Indo-Aryan invaders arrived. The invaders adopted the religion and added their own customs, developing what is known as the **caste system**. Under the caste system, an individual is born into a certain caste (the caste of his or her parents) and cannot move from one caste to another during his or her lifetime. Instead, he or she must remain in the same caste, even in the lowest caste of all (people known as "untouchables"). If a person carries out his or her role well, then he or she will have good **karma** (a good future based on present behavior) and as a result, he or she will be rewarded with a better caste in the next life. If a person does not carry out his or her role well, he or she will have bad karma, and he or she will be punished with a worse caste in the next life.

> The Hindu notion of karma exists in Western culture today. Have you ever heard the phrase, "What comes around goes around"?

Obviously, this system created social stability like few systems can. Few people challenged the ruling class because to do so would mean an even worse lot the next time around. Nearly everyone made a contribution to society based on the caste into which he or she was born, and children followed directly in parents' footsteps. Social life under the caste system, therefore, was very predictable. (The caste system was outlawed in 1949, but hundreds of millions of people in India still follow it as part of their traditional custom).

Buddhism: You can be happy, too

In 260 BC, during the Mauryan Dynasty, Ashoka (the Mauryan king) converted from Hinduism to Buddhism. Buddhism started when **Siddhartha Gautama** (Buddha) criticized the caste system after obtaining what he called "enlightenment." Under Ashoka, missionaries were sent throughout Asia spreading the new religion in the hopes of supplanting Hinduism. Central to Buddhism are the **Four Noble Truths**: 1) Life in this world involves unhappiness; 2) unhappiness is caused by the desire of worldly things; 3) happiness can be achieved by detaching oneself from these worldly things; and 4) nirvana can be reached by those who follow the **Eightfold Path** (basic rules of conduct and thought).

In other words, Buddhism is very thought- and meditation-based. Any individual can reach **nirvana** (peace and happiness), regardless of the station of one's birth. This new religion was an enormous challenge to Hinduism and the caste system. However, after the death of the last Mauryan king in 185 BC, Buddhism slowly disintegrated in India and was replaced by a revival of Hinduism. However, Buddhism thrived in China, Japan, and Southeast Asia, where it was spread by Ashoka's missionaries. Today, Buddhism still has millions of adherents in these regions, while India is still mainly Hindu.

The Gupta Dynasty: India gets a piece of the pi

Nearly all great civilizations have golden ages. To an extent, that's what makes the civilization so "great." During the Gupta Dynasty (AD 320–550), the Indian subcontinent enjoyed a golden age that rivaled that of Athens and Rome. Although the Gupta civilization excelled in many areas of scholarship and the arts, its most significant global contribution was in the field of mathematics. Indian mathematicians developed the decimal system, the concept of zero, the system of numerals that later became known as Arabic numerals, and the concept of *pi* (the ratio of the diameter of a circle to its circumference). The mathematical concepts spread west into the Middle East and from there to the rest of the world.

After the fall of the Gupta Dynasty in AD 720, northern India was invaded by **Moslem Turks** who brought a third major religion, Islam, into the region. While Islam greatly effected the northern Indian subcontinent, Hinduism remained strong in the South. The division created by this mixture of Islam and Hinduism is still felt on the Indian subcontinent today.

Quick Quiz 7

1 **Under the caste system, an individual**

    A can move up the social ladder during this life if he or she works hard

    B is born into a particular caste based on luck

    C must live according to the duties of the caste into which he or she is born

    D may be asked to join a higher a caste if he or she pleases the people in
        the higher castes

2 **The idea that an individual, regardless of caste, can reach nirvana through the Four Noble Truths and the Eightfold Path is associated with**

    F Hinduism

    G the Caste System

    H Buddhism

    J the Gupta Dynasty

3 The Gupta Dynasty is best known for contributions in the field of

    A mathematics

    B textiles

    C drama

    D medicine

Check your answers to Quick Quiz 7 on page 70. If you missed any of these questions, reread the previous section.

China

Building a wall where geography fails

Classical Chinese civilization was centered around the Hwang Ho River valley. This valley was perfect for the development of a civilization, not only because of the agricultural opportunities but also because it had so many natural protections against invasions from the outside world. Deserts lie to the West and North, and the Himalayan Mountains lie to the West and South. The only problem was the Manchurian Plain, which was located directly to the north of the river valley. This plain was China's greatest geographical worry because invaders could easily cross it. To stop invasions, the Chinese (beginning in the third century BC) built an enormous wall out of earth and stone, known as the **Great Wall of China**. For centuries, the Great Wall was an attempt to defend against raids on northern cities. Like all lines of defense, sometimes it worked, sometimes it didn't.

Still, though, China didn't want to remain totally isolated. For the purposes of trade with the West, the **Silk Road** was built, extending westward out of China, south of the Gobi Desert, and into Eurasia, where it hooked up to roads going all the way to Rome. China was known in the West for its fine silk. Hence, the name "Silk Road."

Confucious: An orderly life is a life worth living

Between China's first civilization along the Hwang Ho River valley (Shang China) and the T'ang Dynasty (on which the test writers will focus), traditional religious, philosophical, and social customs became highly developed in Chinese culture. During the Zhou Dynasty (China's longest-lasting dynasty), **Confucius** and **Lao-tzu** had a huge impact on the social structure of the region. Confucianism was and remains a life philosophy more than a religion because it was concerned with daily conduct in the here and now, not with salvation or the afterlife. Confucius focused on individual relationships—the duties that individuals owe to one another within the family and their interactions with superiors or inferiors. Confucian philosophy assumed a class-based society wherein inferior classes owe respect to superior classes, and superior classes (and most importantly the ruling class) owe fairness and kindness to the inferior classes.

Confucianism affected Chinese culture enormously. For centuries, it reinforced the class system and repressed new ideas from the younger generations. It also contributed to internal order within the society. As long as everyone played out their expected roles, society was orderly and predictable.

Lao-tzu stressed inner harmony and peace (similar to Buddhism) in **Taoism**, a religion that claims that happiness and wholeness can be achieved through contemplation, meditation, and the indifference to material things. Taoists believe that the entire universe and everything in it is part of a single whole. The creator, known as the Tao, does not have a particular identity but rather is the amalgamation of everything—good and evil, light and darkness, love and hate. Taoists refer to these pairs of opposites as Yin and Yang.

While Confucianism emphasized the role of class in Chinese society, Taoism emphasized personal spiritual happiness.

Like Confucianism, Taoism contributed to the social and political order of traditional China. For thousands of years, Chinese rulers were considered divine, acting from a **Mandate of Heaven** (similar to the Divine Right Theory, which developed later in Europe). This Mandate of Heaven, however, did not justify absolute power and tyranny. Instead, the mandate was good only as long as leaders ruled justly. As soon as they began to rule unjustly, it was determined that they lost the mandate. (This concept of responsible leadership traces its roots to Confucianism, of course).

The golden age of the T'ang Dynasty

During the seventh to ninth centuries AD, the T'ang Dynasty in China was accomplished in virtually every category of human endeavor—art, architecture, science, philosophy, porcelain-making, silk-weaving, construction of transportation systems, and more. Yet, it is probably the poetry of this period that made the T'ang Dynasty unique. Today it tells us the story of life in China during that time. China developed printing processes, and so its literature spread throughout Asia, influencing the development of literature in Korea and Japan. The T'ang Dynasty also increased educational opportunities for its people and created civil service examinations for government positions.

1   **All of the following religions or philosophies are associated with classical China EXCEPT**

    A Buddhism

    B Hinduism

    C Taoism

    D Confucianism

2   **The Silk Road was primarily built in order to**

    F serve as an escape route for Chinese from invading tribes

    G allow T'ang city-dwellers easy access to farmlands

    H make it easier for Buddhists to make pilgrimages to holy Buddhist sites

    J provide a link between China and the West for the trade purposes

3   **The Mandate of Heaven meant that**

    A Chinese rulers could do whatever they pleased.

    B Chinese rulers could act as leaders of the church, but not as leaders over the government.

    C Chinese rulers were considered divine, but they were considered to have fallen out of favor with the gods if they acted unjustly.

    D Chinese rulers considered themselves to be at the same social level as the peasants.

4   **Which of the following puts three important ancient Chinese civilizations in the correct order, from oldest to most recent?**

    F T'ang, Zhou, Shang

    G Shang, T'ang, Zhou

    H Zhou, Shang, T'ang

    J Shang, Zhou, T'ang

Check your answers to Quick Quiz 8 on page 70. If you missed any of these questions, reread the previous section.

Japanese Geography: A World of Its Own

Since Japan consists of four main islands off the coast of mainland Asia, it was relatively isolated for thousands of years. Ideas, religions, and material goods traveled between Japan and the rest of Asia, but the rate of exchange was relatively controlled. The result was that Japan developed its own unique culture. (Only recently has Japan been overrun by influences from the outside world, but even today, Japan remains extremely distinct.)

Religions: Two for the price of one

Shintoism, a religion unique to Japan, developed more than 2,000 years ago in Japan. Central to Shinto is the *kami*, a term that refers to anything that has the power to support life, including gods, ancestors, and even parts of nature and weather patterns. The goal is to become a part of the *kami* by following certain rituals and customs. The religion encourages obedience and proper behavior.

Another belief of Shintoism is that the gods created the Japanese islands as a source for all other nations and as superior to all other nations. The emperor and his family were considered direct descendants of gods, and therefore divine.

Shintoism was not the only religion to influence traditional Japan, however. Buddhism and Confucianism were introduced from China. Interestingly, many Japanese identified themselves with *both* Buddhism *and* Shintoism. In other words, unlike most other people in the world who only identify themselves with one religion at a time, many Japanese followed two religions simultaneously (and to this day many Japanese manage to balance the influence of both religions in their lives).

Kush and Ghana in Africa

Early African civilizations: Trade

The most significant early civilizations in Africa were the Egyptians and the Carthaginians, both of whom were discussed earlier in this chapter. Both of those civilizations were located in North Africa along the Mediterranean. Other civilizations in sub-Saharan Africa during the same period as the Egyptians included Nubia, Axum, Nok, and Bantu cultures, but it is unlikely that the test writers will ask about any of these civilizations. Instead, they will focus on two civilizations which flourished: Kush and Ghana.

When it comes to Kush and Ghana, the test writers will focus on one thing and one thing only: trade. Kush was located in East Africa, so its most important trading partner was Egypt. Ghana was located in West Africa, so trading was a considerable feat since the traders had to travel tremendous distances just to reach other civilizations.

Learn more about trade in Africa and how it has been affected historically by the Sahara Desert, in chapter 6.

The significance of trade is that it brought the African interior, which for much of African history was totally isolated from the rest of the world, into contact with other peoples and customs. With the trade routes from sub-Saharan Africa linked to the Middle East, the stage was set for future contact between the two regions. Sub-Saharan Africa really took off after AD 1000 with the Mali and Songhai civilizations (and those civilizations will be on the *other* SOL world history test!)

Mayan Civilization in the Americas

From about 300 BC to AD 800, Mayan civilization dominated present-day southern Mexico and parts of Central America. The Mayan civilization was similar to many other civilizations at the time in that it was a collection of city-states, but all the city-states were ruled by the same king. Interestingly, like the Egyptians, the Mayans were pyramid-builders and also wrote using hieroglyphs. The golden age of Mayan civilization (during the sixth to ninth centuries) produced many great works of scholarship and a complex calendar system, but we know most about its architecture and city-building, since remains are discovered still. The Mayans built tremendous cities populated by more than 100,000 people, complete with public plazas, monuments, and pyramids.

Quick Quiz 9

1 **With which two religions did most people in traditional Japan identify?**

   A Buddhism and Hinduism

   B Buddhism and Shintoism

   C Shintoism and Hinduism

   D Buddhism and Christianity

2 **Ancient Kush and Ghana in Africa were significant to world history because**

   F They were the birthplace of Islam.

   G They traded with the Middle East and opened up isolated parts of Africa to the rest of the world.

   H They developed a complex calendar system that became the model for the modern Western calendar.

   J They are generally recognized as the first democracies.

3 **The ancient Mayans and Egyptians had which of the following in common?**

   A They both used hieroglyphs and built pyramids.

   B They both were river-valley civilizations.

   C They both were located in Africa.

   D They both developed an alphabet and a system of numerals.

Check your answers to Quick Quiz 9 on page 70. If you missed any of these questions, reread the previous section.

History Skills

Don't freak out about the history skills questions. In terms of content, the history skills questions will cover pretty much the same information you just reviewed. Skills questions will presume you know the basic facts, plus ask you to draw some conclusions based on those facts.

The history skills questions are going to fall into three basic categories:

1. Generalizations

2. Population distribution

3. Political boundaries

It probably seems like some of these categories should be geography skills questions instead of history skills questions! But that's not how the test writers break things down. In the end, it doesn't matter what you call these questions, it just matters that you're going to have eight of them. So make sure you understand how to apply the historical information you just reviewed in the context of a history skills question.

Generalizations

The test writers will include several questions on the exam that will involve a quotation or an excerpt from a primary source or a secondary source. A primary source is a person who is directly involved in the historical situation. For example, a quote from Julius Caesar about life in Rome during the Roman Republic would be a primary source. So would a quote from a slave living in Rome. A secondary source is a book or a person or a point of view that was not involved directly in a given situation. For example, a modern textbook about ancient history is a secondary source. None of the authors of the textbook lived in ancient civilizations—they're simply passing along information about events. The book you're reading right now, for instance, is a secondary source. The test writers may ask you to identify whether a source is a primary source or a secondary source, so make sure you understand the difference. More often, however, they will use sources in questions and ask you to make generalizations from the source.

For example, look at the following excerpt from the Code of Hammurabi:

If a seignior [noble] has knocked out the tooth of a seignior of his own rank, they shall knock out his tooth. But if he has knocked out a commoner's tooth, he shall pay one-third mina of silver.

The test writers will ask you something like:

Based on the passage above, what generalization can you make about Babylonian society?

And you would (or should!) pick the answer that says something like:

Divisions existed between the social classes.

Population Distribution

The test writers will want to know if you understand how history has been impacted by population distribution and how population distribution has impacted history. Now don't panic! You don't have to sit around and memorize the populations of cities and kingdoms and empires. You just need to have a general understanding of population distribution, which you should already know from the previous four sections in this chapter.

These are the main facts about population distribution to keep in mind for this test:

- Before the rise of the early civilizations, populations were nomadic, constantly moving from one grazing land to the next based on the needs of the herds.

- Early civilizations were located in river valleys. The populations clustered themselves along the river's banks on farms and in towns and cities as the farms began to generate large food surpluses.

- In the ancient world, the highest concentrations of people were in the Mediterranean region, the Indus Valley, the Hwang Ho Valley in China, and increasingly in Central America.

- As empires grew, wealth grew, and this meant that an increasing number of people could live in cities. Thebes, Babylon, Athens, Rome, Carthage, and, later, Constantinople and Baghdad are examples of cities that grew dramatically as their empires grew.

- During the European Middle Ages, the vast majority of the population was **agrarian** (meaning they worked on farms or in farming villages) as part of a fief.

- As trade increased between civilizations, the populations distributed themselves in port cities (like Athens and Constantinople) and along trade routes (like Samarkand and other cities along the Silk Road).

- Be aware that, in general, the arts and sciences have been advanced in civilizations in which there have been large populations to support and raise the money for them.

- Never forget that populations generally go toward arable land (which often means river valleys) and jobs (which often means cities that are the centers of trade in major sea harbors or along land trade routes).

Political Boundaries

Political boundary questions will test your history knowledge with respect to contemporary political geography. In other words, the test writers will want to know if you understand the current locations of certain historical empires.

Some ancient empires are easy: ancient Egypt was located approximately where modern Egypt is located; ancient Greece was located where modern Greece is located; and ancient India was located where India and Pakistan are located today. But some empires are not so obvious, and these are the ones the test writers will focus on.

Here's a list of some important ones:

- Mesopotamia is now part of the country of Iraq.

- Mayan civilization occupied territories that are currently part of Mexico.

- The ancient Roman territory of Gaul is modern-day France.

- The heart of the Byzantine Empire is modern-day Turkey, and the city of Constantinople is now known as Istanbul.

- Mecca and Medina, two of the most important cities in Islam, are located in modern-day Saudi Arabia.

Quick Quiz 10

1 Which of the following is NOT a primary source?

  A a handwritten journal about the Battle of Tours written by a soldier who fought in it for a person who is studying the Battle of Tours

  B the original manuscript of *The Politics* by Aristotle for a person who is studying Greek political philosophy

  C a copy of the Qu'ran for a person who is studying the basics of the Islamic faith

  D an article written by an eminent scholar in a well-known historical journal about the Peloponnesian War for a student who is studying the impact of the Peloponnesian War

2 During the European Middle Ages, the vast majority of the population was

  F agrarian and living on fiefs

  G living in cities

  H nomadic

  J employed in the commerce of trade

3 Mesopotamia is in modern-day

  A Iraq

  B Israel

  C Egypt

  D India

Check your answers to Quick Quiz 10 on page 70. If you missed any of these questions, reread the previous section.

Answers to Quizzes

Initial Practice Quiz

1 C	2 F	3 D	4 H	5 A	6 H	7 C	8 G
9 C	10 J	11 C	12 J	13 B	14 G	15 A	

Quick Quiz 1: 1 C   2 F   3 B

Quick Quiz 2: 1 A   2 J   3 B   4 J

Quick Quiz 3: 1 A   2 H   3 C   4 A

Quick Quiz 4: 1 A   2 H   3 D   4 J   5 D   6 F   7 C

Quick Quiz 5: 1 B   2 G   3 B   4 G

Quick Quiz 6: 1 A   2 H   3 D   4 G

Quick Quiz 7: 1 C   2 H   3 A

Quick Quiz 8: 1 B   2 J   3 C   4 J

Quick Quiz 9: 1 B   2 G   3 A

Quick Quiz 10: 1 D   2 F   3 A

Chapter 4

Review of Geography through AD 1000

What You Need to Know

First things first: Review chapter 3 before reviewing this chapter. The geographical information in this chapter is dependent on historical context. The geography questions will expect you to look at historical facts and apply the correct geographical concept to the question. If you understand the historical facts in chapter 3, you'll get more from this chapter than if you don't.

Keep in mind that you're not expected to know everything from your tenth-grade geography class. Some of the stuff from your geography class will be tested on the World History and Geography to AD 1000 exam, some will be tested on the World History and Geography from AD 1000 exam, and some won't be tested at all. For the purposes of the World History and Geography to AD 1000 exam, it makes sense to focus on the impact of geography through AD 1000. This chapter is designed to help you focus on the right stuff.

You should have already studied for these categories in chapter 3:

1	Ancient civilizations	7	questions
2	Greece and Rome	11	questions
3	Middle East, Russia, and early medieval Europe	9	questions
4	Asia, Africa, and the Americas	8	questions
5	History skills	8	questions

Now, in this chapter, you'll also study two new big categories:

6	Geography knowledge and concepts	12	questions
7	Geography skills	6	questions

Interestingly, the geography questions on this exam won't ask, "What was the capital of the Byzantine Empire?" You *do* have to know the answer to that question (the answer is Constantinople), but the test writers consider that a history question, *not* a geography question. Geography questions ask about the impact of humans on the land, or the impact of the land on humans. They also ask about the characteristics of maps and mapping, and about the distribution of cultural characteristics throughout the world (especially due to migration). In other words, geography questions ask you to apply general historical information to an understanding of geographical concepts and skills.

> Read chapter 3 before you read this chapter so that you will be armed with geography skills *and* historical context.

How to Study

You may want to take a few moments to flip through this chapter to get a feel for what you need to know. If a lot of this stuff looks new to you, spend plenty of time reading and learning the details (the topics covered in this chapter will account for 18 questions on the exam). If, on the other hand, you recall a lot of this information from your geography class, you'll be ready for the exam after reading through this chapter once.

In the end, you should be able to piece together enough points from enough categories of questions to reach your goal.

Geography Knowledge and Concepts

There will be 12 geography knowledge and concepts questions on the exam in the following categories:

A Interaction between humans and their environments

B Characteristics of geographical regions

C Cultural characteristics of the landscape

D Migration and cultural interaction

E Patterns of urban development

> If geography skills questions bug you, then study geography knowledge and concepts questions to make up for it.

Interactions Between Humans and Their Environments

Where to put all the water?

One aspect of life on Earth has troubled people since the beginning of time. There's plenty of water, but too often it's not where we want it to be. Either there's too much of it in a given location (like a flood) or not enough (like a drought), or we're using the land in such a way that we want to control the rate at which water becomes available. On this exam, you'll answer questions about how early civilizations dealt with water.

There are a few things to focus on. First, since the early civilizations were in river valleys, they had to deal with flooding. The test writers will likely focus on the Nile River Valley, but they might also ask about the Euphrates.

The great thing about floods is that after the floodwaters recede, a new layer of rich topsoil is left behind, making the land prime for farming. The early river civilizations, then, almost constantly had access to replenished farmland, so they could farm on the same property for centuries. Without this renewal, the farmland eventually would be stripped of its nutrients, leaving it unsuitable for the growth of major crops. Early civilizations became sophisticated engineers of flood control. They could predict when floods would occur and try to divert waters into farmlands and away from towns.

To get water where they wanted it to be, ancient civilizations used a series of aqueducts. At first, these were simply a series of ditches through which water flowed. By the time of the Romans, aqueducts were very sophisticated. Some were elevated, or made of stone, and carried water over gorges, valleys, streets, and anything else that was between where the water was and where it needed to be.

Landscape affected where people settled

Would you want to live in a place where the temperature never climbed above zero? Or a place where there isn't much food or water?

What do San Francisco, New Orleans, and Chicago have in common? They're all big cities. They also are all near large bodies of water, which is a major reason why they're all so big.

Landscape always has had an enormous impact on human beings. Early civilizations like ancient Egypt, Mesopotamia, and Shang China all thrived along rivers because river valleys provided what was needed. Rome and Athens thrived because their landscapes provided protection from military invasions and also strategic positions for economic opportunities. It just makes sense that the landscape determines, in part, where people decide to settle.

People also affect landscape

Think about a nice river valley. The water flows past trees and rocks and wildlife of all sorts. Birds chirp. The plants are overgrown and tangled together, some of them dangling in the water of the river.

Now, imagine a gigantic city on the river. Sewage spills into the water. Boats of all kinds carry people and products and raw materials to and from shipping ports. Buildings line the shore. Birds perch on top of statues and turrets instead of trees.

Now, think about both the clean river valley *and* the city in the exact same location (but just at different times), and you have an understanding of just how much humans impact the environment. For every city that exists, there was once a piece of natural, uninhabited land.

Early civilizations may not have affected the landscape to the degree that modern civilizations have, but even then, the impact of humans on the environment was staggering. Early civilizations settled in patterns such that they could use river systems to their advantage. The population, therefore, was clustered along river banks, which changed the landscape substantially. If building materials weren't available near the water, the people brought them from someplace else (stone was moved by the Egyptians from outlying areas into the river communities, for example). This meant that the Earth was (and is) affected by humans just as the Earth affect humans.

The real impact on the Earth came with the Agricultural Revolution. When people learned to farm, they started building stable societies and stopped moving around as much. Nomadic people didn't change the land nearly as much. They responded to the land rather than trying to change it. But farmers truly changed the land. Not only did they clear land to make it suitable for farming, but they introduced new plants that had not grown there previously, moved water around using irrigation, and made it possible to develop towns and cities by growing enough food not just for themselves, but for others. Once it got to the point where they needed only one farmer to feed every five or ten people, that farmer could farm and the others could do something else and still have food to eat! So they built cities. Which means that the Agricultural Revolution didn't just change the land that was farmed—it changed everything.

> With time on their hands and enough food to live, people were able to build towns and cities.

Managing time

The relationship between humans and our planet is so strong that this relationship is reflected even in our concept of time. Our hours, days, months, and years are based on the Earth's spin, the Earth's orbit around the sun, and the moon's orbit around Earth. Our seasons exist because of the Earth's tilt.

Interestingly, however, human activities have also impacted the calendar. In the Western world, we use a calendar system based on the birth of Jesus Christ. In China, the calendar system is based on the beginning of recorded Chinese civilization. The ancient calendar system of the Maya was based on 5,200-year cycles of destruction and rebirth.

Characteristics of Geographic Regions

What are regions, anyway?

A region is a name we give to a place that has something different or distinctive about it when compared to other places. Regions usually don't have definitive borders. Instead, they're general locations in which some pattern dominates.

Imagine saying something like, "I live in a warm region of the country." Or a "rainy region," or a "conservative region," or a "football-obsessed region." The "warm region" doesn't have exact boundaries. It's not like the weather is warm only within certain borders and then suddenly becomes very cold. You don't cross a bridge and a sign tells you that you've now entered the "warm region"! It's just a name we use for a general location. Nobody's really sure where it begins and ends.

We use regions when we talk about history and politics all the time. For the purposes of this test, you'll be expected to understand the difference between regional names based on the land and regional names based on cultural characteristics.

> Many of the political and cultural differences that exist between countries today developed because of geographic differences.

Physical regions are based on the land

There are two major physical regions that the test writers will expect you to know about and understand why they are called physical regions in the first place. The first physical region is Mesopotamia and the second is the Mediterranean world.

Mesopotamia is the name given to the land between the Tigris and Euphrates Rivers, or more precisely, between and including those river valleys. In other words, a person who lives between the two rivers doesn't look across the Euphrates and claim that those people aren't in Mesopotamia. They all are. And where the valley begins and ends isn't always agreed upon. It's kind of like saying you're going to "the coast." Do you mean the actual spot where the water ends and sand or rocks begin? (And if so, do you mean at high tide or low tide?) Do you mean the whole beach? Do you mean the towns on the beach? How far inland are you willing to go and still call it "the coast"?

It's the same deal with Mesopotamia. Historians refer to Mesopotamia in general, and they include the civilizations along the rivers. But there isn't a formal line that you cross. Ur was clearly in Mesopotamia. Damascus was clearly not. In between, well, you can decide where it begins and ends. Historians usually try to respect self-determination. If a town or group of people identified themselves with Mesopotamian culture, then the town would likely be included in the Mesopotamian region. Still, though, it's called a physical region (as opposed to a cultural region) since the *main* feature of Mesopotamia is a physical characteristic.

The other common physical region that you'll be tested on is the Mediterranean world. This region is the land around the Mediterranean Sea. Where it starts and ends is anyone's guess, but it is clear that it is a physical region and not a cultural region. The Mediterranean world is very diverse: Christians on the north side, Muslims on the east and south side, Jews sprinkled in between. The Mediterranean world encompassed lots of different races and ethnic groups. Some of these groups were friends and some were enemies. It's definitely not a cultural region since the region isn't united under one culture.

The Appalachians are another example of a place that is distinguished by a geographical feature—the mountains—but encompasses the regions around the mountains as well. Those regions are called Appalachia.

The Mediterranean world is so widely recognized as a region that historians usually don't say the "world" part. Instead, they just say "Mediterranean." They don't mean the sea itself, but the region around the sea. (If you say, "He grew up in the Mediterranean before moving to sub-Saharan Africa," you'd mean he grew up in the Mediterranean world, not in the sea itself.) The bottom line is that the region identifies itself with the sea, which is a *physical* entity, and so it is a physical region.

Cultural regions are based on human cultures

Cultural regions share a common cultural identity without regard to the characteristics of the physical landscape.

For the test, you'll need to know about a few historical cultural regions. The Roman world, Alexander the Great's empire, the Islamic world, the Byzantine Empire, and the Silk Road are all examples of regions that have no definite boundaries (although we do our best to draw lines on maps for these regions anyway!). But they all have common cultural characteristics specific to each region. The Roman world, for example, was a cultural region that extended not only throughout the Mediterranean (which is a physical region) but also into Gaul (which is present-day France) and Britain (which is an island). Why was it all part of one cultural region? Because the entire Roman world shared cultural characteristics that the Romans brought with them when they conquered these physical regions. Since the entire Roman world had shared cultural characteristics, Londinium (now London) was part of the same cultural region as Jerusalem!

Sometimes we give labels to cultural regions based on their main characteristic. For example, Alexander the Great's empire was largely an *economic* region because the locations in the empire were linked by trade. Often, politics, economics, religion, and culture all go hand in hand. The Islamic Empire, for example, shared a religion and a language, and as a result developed similar cultural institutions and social structures. What's more, the cities within the empire traded with each other and offered each other military support, so it was an economic and allied region as well. The region is so culturally unified it even shares a common architectural style!

Finally, you should realize that one city or location can belong to a lot of regions throughout history, and can belong to several regions simultaneously. Jerusalem, for example, historically has been part of several cultural regions—the Hebrew culture, the Roman Empire, the Byzantine Empire, and the Islamic Empire—as well as several physical regions—the Mediterranean world and the Fertile Crescent.

> Remember that regions can overlap each other, and their boundaries are hard to define.

Cultural Characteristics of the Landscape

Just by looking at a location or a city, often you can tell the cultural characteristics of the place. Sometimes you can even tell that certain historical events have occurred. Just by looking around Virginia, for example, you might realize that Virginia was important during colonial times. You'd be able to tell by the architecture, the names of towns and universities, and by road signs announcing historical sites. You'd also be able to tell that the state played a significant role in the Civil War. There are remnants of Civil War heritage everywhere.

The same is true all over the globe. If a culture bothered to build something somewhere, you can infer a lot about that culture's background by studying what was built. On the test, you'll be expected to understand that cultural characteristics are reflected in what is built and in the manner in which it is built.

Religious buildings—like mosques, churches, temples, and pagodas—typically are designed with minute attention to symbolism, history, and message. Christian churches built during the Middle Ages, for example, typically have the altar at the intersection of a long nave and two side galleries. In other words, the floor plan is in the shape of a cross. Often the windows or walls are decorated with figures from Biblical stories.

In contrast, mosques in the Islamic world are beautifully decorated with geometric patterns but devoid of any representations of the human figure, since Islam holds that human representations are a form of idolatry and blasphemy. Muslims believe that only God can create the human body and that attempts to reproduce it are, therefore, attempts to think of oneself as a god.

Civic buildings, as well, give us ideas of the cultures that produced them. The Colosseum in Rome, for example, would lead any observer to conclude that the Romans enjoyed large-scale entertainment and sport, both because of the impressive size and beauty of the arena, and because of its prime location in the historical city.

Furthermore, walls and barriers also tell us a lot about past cultures. The Great Wall of China, for example, not only tells about Chinese vulnerability and the direction from which they feared attacks, but also about the characteristics of the Chinese social structure. Only an organized society could build such a massive wall.

A monument tells a lot about who and what a culture honors, and how they do so.

Finally, statues and monuments very literally tell us about the history of a location. Statues and monuments tell us details about the past event or person and they also tell us that the society that built it thought it was an important person or event to commemorate.

Migration and Cultural Interaction

Historians and geographers talk about migration a lot because migration patterns explain so much about our world and our world's population. A migration is a permanent move to a new location. Traveling around the world or staying a few months at your aunt's house in Nebraska don't count as migrations. Picking up and leaving your home so that you can establish residence someplace else *does* count. It counts even if you don't know where you're going to end up. You migrate when you leave a place with every intention of staying somewhere else. It doesn't matter how long you stay.

There are two broad categories of factors that cause the human population to migrate. These two categories are called *push factors* and *pull factors*. A push factor is something that pushes you away from a given place. A pull factor is something that pulls you closer to a given place. Unless you're aimlessly wandering around the countryside for months at a time, it's likely that when you move, you do it for one of these two reasons.

Push factors: Get me out of this place!

Push factors are behind a great deal of human history. When people are pushed out of a place, they often don't even know where they're going to end up. This has made our world very unpredictable. A lot of places are extremely diverse because different kinds of people were pushed together for different reasons.

For the purposes of the test, the most common push factors until AD 1000 were 1) over-population, 2) religious persecution, 3) agricultural decline or famine, 4) war and conflict, 5) slavery, and 6) natural disasters, like earthquakes and floods. If you think about these, you will see that they all make a lot of sense as major push factors.

Together, these six factors are responsible for moving millions and millions of people, sometimes in groups no larger than a family or two and other times by the hundreds of thousands. These six factors are responsible for spreading out the population into some really bizarre places where you wouldn't expect to find people living. It's sad but true. Sometimes people move to places that aren't all that great simply because the place they came from was even worse. They then have kids, and their kids think of it as home, so the people stay. Other times, people are pushed from one location and find that their lives greatly improve in their new environments.

There are an endless number of examples of push factors from history that the test writers can use. They might ask about the persecution of Jews that resulted in Jews moving out of Palestine and into Europe, Africa, and other parts of Asia. They might refer to an earthquake in Greece or a war in Africa or rivers drying up in India. In all cases, you won't need to know the details. Once the test writers start asking about push factors, you'll know that the correct answer has to be something bad that's happened.

> If your family moves from one place to another because there are no more jobs in your old town, you moved because of a push factor.

Pull factors: I can't wait to get there!

Sometimes people migrate from a place not so much because they don't like where they're living but rather because they *really* like the place they want to go, or they really like what they've heard about it (sometimes they're disappointed when they get there!). These things that they really like are pull factors.

The biggest examples of pull factors before AD 1000 were 1) arable land and 2) religious acceptance. Economic opportunity, of course, is a huge pull factor as well, but the test writers likely won't test that concept on the pre-AD 1000 exam. Whether the test writers ask about river valleys pulling in nomadic people and turning them into farmers or Christians flooding into Constantinople, you will know that if they refer to a historical event that made people want to move to a specific place, they're asking about a pull factor.

Keep in mind that push factors and pull factors can work simultaneously. If you're being persecuted in Persia, you'll want to leave. If you hear that you won't be persecuted in Constantinople, then that's where you'll want to go.

Impact of migrations

Migrations have spread languages, religions, and social customs to all parts of the world. Migrations have led to the expansion of some cultural regions (such as the Islamic Empire) and the fall of others (such as the early Indus valley civilizations). Migrations have spread architectural and cultural characteristics to new parts of the world (such as Islamic architecture in Spain), and philosophical, political, and legal systems throughout huge regions (such as Roman law). The list goes on and on.

> Because early peoples moved around, more land was inhabited, which led to technological and agricultural advances and cultural diversity.

The bottom line is that if everybody stayed in the same place throughout history, most of the world would be uninhabited, the river valleys would be packed, and each civilization's culture would be unique. Instead, people have been on the move constantly, and the amount of cultural diffusion between migrants has been enormous.

Patterns of Urban Development

When you think about why towns, cities, and farms happen to pop up where they do, you're thinking about the patterns of urban development.

You already know that the first major urban development pattern occurred because of river valleys. Mesopotamia, ancient Egypt, early Indus civilizations, and Shang China were all in river valleys. And yet all these places developed independently. Coincidence? Probably not. More likely, people are just plain smart. When faced with choices and good information, most people follow a predicable, logical pattern. Personal experience showed early civilizations that river valleys were practical places to farm. River valleys also provided a method of transportation and a steady supply of water.

The river valley example isn't the only pattern, however. Towns and cities are located in certain places for a variety of reasons, but the patterns are very clear when you look at enough cities.

Geographers group patterns of urban development into two categories: *site factors* and *situation factors*. Site factors are specific things about a particular location that make it desirable, without regard to what's around it. For example, the natural harbors are good, the resources are plentiful, and the weather is nice. Situation factors, on the other hand, aren't about the things that make a place so good all by itself. Instead, situation factors are the things that make a place attractive *in relation to other places*. For example, your friends live close by and you don't have to drive far to school or work.

If you say you like to sit in a chair because it is comfortable and it looks out over the ocean, then it's a *site* factor. If you say you like a chair because of it's relation to something else (for example, when you sit in the chair, the teacher can't see you!), then it's a *situation* factor. The fact that you sit there has little to do with the chair (it might actually be uncomfortable) and everything to do with the teacher. If the teacher were to move to a position in eye-shot of the chair, you wouldn't like the chair as much.

Site and situation reflect the two great patterns of urban development. Cities tend to be established either because the site is excellent or because the situation relative to other things is excellent, or both.

Site factors

The locations of early river-valley civilizations are examples of the effects of site factors. It was the specific nature of the land in the river valleys that led to the building of the civilizations. Another good example is ancient Athens. Athens grew where it did because of the particular factors of that precise location. It grew on a peninsula jutting out into the sea, where the harbors were great and the view was spectacular. Indeed, Athens is also a great example of situation since it became a major civilization in large part because of trade and cultural exchange with other cities.

Situation Factors

Perhaps one of the best examples of a situation factor was the selection of Constantinople as the new capital of the eastern Roman Empire (Byzantine Empire). The location wasn't chosen so much because of the characteristics of the site, but rather because of its situation (although the site is quite spectacular as well). The main reason it was selected is that it was well protected from enemies. It was farther east than Rome and therefore more secure against invasions from the north and west. And it was more centrally located to burgeoning Christian communities. So it was its location relative to other cities and civilizations that made it a wise choice. Other examples of cities that thrived because of their situations are Baghdad, Damascus, Mecca, and cities like Samarkand that grew up along the Silk Road. All of these cities grew because of their location on trade routes. Therefore it was their situation relative to other places that made them so ideal.

> You'll learn more about how location affects a country's military and cultural strength in chapter 6.

Geography Skills

There will be six geography skills questions on the exam.

Geography skills questions are all about maps. The questions, however, don't ask you to find a city on the map. Instead, they ask you if you know *about* maps: how they're made, how they're organized, and what they mean.

You'll be expected to understand five different things about maps:

1. Map projections and distortion

2. Scales

3. Latitude and longitude

4. Points of view

5. Thematic maps

Map projection and distortion

Maps are, as you know, flat sheets of paper that you can hang on a wall or unfold in your car. But the Earth, which is what maps represent, is a sphere, *not* a flat surface. When we try to put a sphere on a flat piece of paper, we run into some problems. A sphere doesn't nicely flatten onto a rectangular map. We can stretch and pull and cut and reconfigure, but in the end, some places on the flat map will look closer together and smaller than they were on the sphere, and other places will look farther apart and larger.

Cartographers (map makers) have been wrestling with this problem for as long as there have been maps. The act of transferring information from the Earth's surface onto a flat map is known as *projection*, and different projections deal with the problem of distortion in different ways. Probably the most popular projection is known as the Mercator projection, which is popular because it keeps the shapes of land masses relatively undistorted and the direction consistent, and manages to do it all on a rectangular map. How? Well, that's the one big problem with the Mercator projection. The land masses near the poles are much, much larger than they should be when compared to the land masses near the equator (the *shape* is accurate, but not the *size*). As a result, countries like Greenland look as large as the entire continent of South America or Africa, when, in fact, Greenland is much smaller than both of them.

> Have you ever seen a map that was made in Russia? On a Russian map, Russia is in the center and North America is off to the side, whereas in American maps, the opposite is true. You will learn more about point of view in the pages that follow.

Scales

A map's *scale* is the relation between a distance on a map and a distance in the real world. For example, on a road map, one inch might represent 50 miles. This means that if on a map the distance between Richmond and Washington, D.C., is 2 inches, then the two cities are, in reality, about 100 miles apart.

Maps that show the entire world have a huge scale ratio. One inch might represent several hundred miles, depending on the size of the map. These maps aren't able to show much detail. Washington, D.C., probably would be marked on such a map because it's a large city and the nation's capital, but Richmond and Norfolk probably would not be. Smaller towns like Charlottesville and Danville certainly would not.

Maps that show a small area, like a detailed map of Old Town Alexandria, will have a much smaller scale and will be able to show a lot more detail, including the locations of specific buildings. One inch may represent only a half-mile or less.

Latitude and longitude

On a globe, location is determined by latitude and longitude. Latitude and longitude lines are part of an imaginary grid that cartographers have drawn over the surface of the globe.

The longitude lines (or meridians) all stretch from the North Pole to the South Pole, going north and south along the surface of the globe. One of these meridians passes through the Royal Observatory in Greenwich, England, and has been agreed upon as the "starting point." In other words, the Greenwich meridian (known as the prime meridian) has been designated as 0 degrees longitude. On the exact opposite side of the globe, another meridian is at 180 degrees longitude. All of the other meridians are between 0 and 180 and are designated as either *east* or *west*, to show whether they lie to the east or the west of the prime meridian. (For example, Prague, Poland, is at almost 15 degrees east longitude.)

Latitude lines (or parallels) are circles drawn east and west around the surface of the globe. The one in the exact center of the globe (where the globe is at its widest, exactly half-way between the North Pole and the South Pole) is called the equator. Like the prime meridian, the equator is the starting point, but for latitude instead of longitude. The ending points for latitude are the poles. The North Pole is at 90 degrees north and the South Pole is at 90 degrees south. All latitudes in the Northern Hemisphere are between 0 degrees and 90 degrees north, and all the latitudes in the Southern Hemisphere are between 0 degrees and 90 degrees south. (Cairo, Egypt, for example, is at almost 30 degrees north latitude.)

<aside>
A latitude and longitude won't help you much if you need directions to someone's house, but they will help if you need to find some place on a large map.
</aside>

So if you put it all together, every place on the globe has an exact position based on where its longitude and latitude lines intersect. For example, St. Petersburg, Russia, is at approximately 30 degrees east longitude and 60 degrees north latitude.

Points of view

Academics, cartographers, and geographers in our modern world usually try to stay objective, to represent the land as accurately and as bias-free as possible. Still, it's hard to remain entirely objective since a map has to take a certain point of view just by being a map. (You can't include everything in the map or else the map would be as big as the world itself.) In the past, cartographers often didn't even try to remain objective; they designed maps to support the point of view of an emperor, religious leader, or merchant.

Maps, because they are flat and typically rectangular, have a center and four edges. What goes in the center? What goes on the edges? Even if the distances between cities are projected correctly, which cities are included? Which mountain ranges and rivers? If it's a world map, should an American cartographer use the English spellings of cities and countries in other parts of the world or the native spellings? Should cultural landmarks be included? If so, according to whose culture? Christians, Jews, and Muslims may disagree about the most important cultural landmarks that should be included in a map of Jerusalem, for example.

Map making can get complicated. Fortunately, these days, cartographers make thousands and thousands of maps for just about anybody's point of view. But in the past, maps were often culturally lopsided—sometimes intentionally, sometimes not.

For example, medieval maps of the world typically placed Jerusalem at the center of the world. This was because medieval Christianity was highly symbolic. The medieval Christian cartographers were representing the idea that Christianity began with Christ, that the crux of Christianity began with Christ's life in Jerusalem, and therefore, everything on a map should emanate from Jerusalem. Jerusalem was often written in enormous and highly decorative letters, despite the fact that Jerusalem was not the largest city in the world.

Some maps have lines on them that show the migrations of peoples. Roman Empire maps, for example, showed movements of Romans into new territories as "expansions" or "growth," while advancements of groups on the northern fringes into the empire were labeled "invasions" by "barbarians."

For the purposes of the test, you will need to understand that maps almost always have a point of view. The test writers may give an example of a historical map and then ask you to determine the point of view or the culture from which it most likely came.

Thematic maps

Maps are used to show not only the lay of the land, political lines, or geographical features like rivers, mountains, and deserts, but also cultural and demographic information. These kinds of maps are known as thematic maps because each one has a certain theme.

The most common thematic maps include maps that show population density, economic activity, distribution of resources, distribution of world languages or religions, or the location of certain ethnic groups. The test writers may give you a thematic map and then ask you to make a conclusion based on the map.

> There are as many kind of maps as there are ways of comparing different regions. But don't worry: The test won't ask you about any kind of map that's too obscure.

Chapter 5

Review of History from
AD 1000 to the Present

First Things First: Let's See What You Know

If you need to, review the Process of Elimination techniques that were covered in chapter 2, then practice using them in the practice quizzes.

Take the following quiz to see what you remember from your history and geography classes. Guess if you have to, using Process of Elimination techniques to narrow down answer choices, just like you'll do on the real test.

1   **All of the following were characteristics of European feudalism EXCEPT**

   A   Peasants and lords had corresponding responsibilities.

   B   Land was inherited based on primogeniture.

   C   Cities were the main centers of activity.

   D   Lords and knights followed the code of chivalry.

2   **During the Middle Ages, the Crusades were**

   F   launched by Rome to defeat Constantinople

   G   launched by European Christians to acquire land occupied by Muslims

   H   launched by Muslims to convert Europeans to Islam

   J   launched by Eastern Orthodox Christians to convert Roman Catholics

3   **The Magna Carta is a document that**

   A   created the modern nation-state of France

   B   led to the Spanish Inquisition

   C   represents the beginning of democracy in England

   D   established the rule of the czars in Russia

4   **During the European Renaissance**

   F   Europe experienced a rebirth of Greek and Roman roots

   G   Russia emerged as the most powerful country in Europe

   H   Leonardo da Vinci led a reform movement within the Catholic Church

   J   The Medici family abandoned the arts and pushed Europe toward Protestantism

5   **Who is associated with the 95 Theses?**

   A   Martin Luther

   B   John Calvin

   C   Huldrych Zwingli

   D   Henry VIII

6   **Which list puts the feudal Japanese classes in the proper order from highest to lowest?**

   F   Shogun, diamyo/samurai, artisans, merchants

   G   Shogun, artisans, merchants, diamyo/samurai

   H   Diamyo/samurai, shogun, merchants, artisans

   J   Diamyo/samurai, shogun, artisans, merchants

7 **Which of the following was a reason that the Europeans went to Africa in search of slave labor?**

   A It was cheaper to use African slaves than natives from the Americas.

   B Disease killed many of the natives in the Americas.

   C Europeans were over-worked in the Americas.

   D Europeans had heard that Africans were very strong.

8 **Which of the following best describes Divine Right Theory?**

   F Monarchs believe their right to rule is given to them directly by God.

   G Monarchs believe their first-born child is entitled to the throne.

   H Monarchs believe they have the right to acquire neighboring empires.

   J Monarchs believe they should rule only as long as the people consent.

9 **Which of the following is true of the French Revolution?**

   A After overtaking the monarchy, the French immediately established a democracy.

   B Very few people were killed at the hands of the revolutionaries.

   C The French Revolution didn't achieve its goals until after Napoleon's reign.

   D The Jacobins wanted to restore peace by slowly moving from a monarchy to a democracy.

10 **In an attempt to modernize Japan during the late 1800s, the leaders of the Meiji government decided to**

   F study Western institutions and technology

   G maintain a policy of isolationism

   H establish close relations with China

   J end the political power of the Buddhists

11 **According to the theory of mercantilism, colonies should be**

   A acquired as markets and sources of raw materials

   B considered an economic burden for the colonial power

   C granted independence as soon as possible

   D encouraged to develop their own industries

12 **Fascism in Europe during the 1920s and 1930s is best described as**

   F a demonstration of capitalism

   G a form of totalitarianism

   H a classless society

   J a system of humanist ideas

13 Under Joseph Stalin, life in the Soviet Union was characterized by

    A an abundance of consumer goods

    B political instability and numerous civil wars

    C support for small family-run farms

    D the use of censorship and the secret police

14 One characteristic of the apartheid that was practiced in South Africa is

    F forced migration of blacks to other nations

    G integration of all races in society

    H an open immigration system

    J segregation of the races

15 Which of the following is associated with Mao Zedong?

    A the Cultural Revolution

    B the Bolshevik Revolution

    C the French Revolution

    D the Industrial Revolution

Score Yourself

Okay, now score yourself using the answers at the end of this chapter on page 153. If you answered more than 12 correctly, you're in great shape. If you answered between 8 and 12 correctly, you need to do a little studying. If you answered fewer than 8 correctly, read this chapter *completely*! All the stuff that's most likely to show up on your test is right here in this chapter. Read it. Then read it again.

> If you understand the stuff we've included in this chapter, then you'll do just fine on test day.

Stuff You're Expected to Know

1   Late medieval Europe: AD 1000   8 questions
    through the Reformation

2   The Age of Discovery   8 questions

3   Sixteenth through nineteenth centuries: 12 questions
    Enlightenment, Absolutism, Reason,
    and the Industrial Revolution

4   Twentieth-century world conflicts   10 questions

5   History skills   7 questions

And two more categories will be covered in chapter 6:

6   Geography knowledge and concepts   12 questions

7   Geography skills   6 questions

You are expected to know and understand a) the basic facts, individuals, groups and concepts from each of the categories; b) the significance of those facts, individuals, groups, and concepts; and c) the "big picture."

> If you review everything in this chapter a few times, you'll be adequately prepared for the 43 questions that come from these five categories.

How to Study

If you have time, read this chapter chronologically, since history builds upon previous events. You'll understand how world civilizations developed if you read the information in order rather than skipping around for random facts. However, if you already have a big-picture understanding of world history and just need to fine-tune a few key points, use the headings and the words printed in bold to guide your study. For example, if you don't remember much about John Locke, scan the chapter until you find information about him. If the Treaty of Versailles escapes your memory, you'll find it in section 4 of this review.

In the end, you need to be familiar with just about everything in this chapter if you expect to do well on the exam. We've tried our best not to waste your time or ours. Everything included in this chapter is here because the state of Virginia has indicated that these topics, people, places, and events are fair game for test questions. What's more, there's a lot of important history that isn't included in this review, for precisely the same reason: the state of Virginia hasn't indicated that it will show up on the test!

> *Remember:* If it's not here, it won't be on the test.

Late Medieval Europe: AD 1000 through the Reformation

There will be eight questions about late medieval Europe on the exam. The early medieval era was covered in your eigth-grade history class. In that class you started to talk about feudalism, the Byzantine Empire, and the Catholic Church in Europe. Your ninth-grade history class covered the same topics but then showed their impact on future events. In other words, the medieval world is where your ninth-grade class *began*, and so that's where this review begins as well. This section then goes on to show how European civilization developed beyond medievalism.

> By reviewing the material, you increase your chances of getting all eight medieval Europe questions correct!

- The state of Europe and the Middle East at about AD 1000

- Conflicts among Eurasian powers

- Emergence of nation-states

- The Renaissance

- The Reformation

The State of Europe and the Middle East at Around AD 1000

Feudalism: My tiny kingdom is bigger than yours

Feudalism is the name given to the social, economic, and political system during the Middle Ages. The economy was mostly based on farming. The people who *owned* the land were wealthy and powerful, while the people who *worked* the land were neither wealthy nor powerful. In fact, they lived a very meager existence.

Here's how feudalism worked: At the top, there was the king. He's the one who had power over an entire territory of land, called his *kingdom*. Beneath him were the nobles, called **vassals**, who in exchange for military service to the king, were given power over sections of the kingdom. In other words, if two kings in neighboring kingdoms got into a fight, the two kings would call up their nobles and go to war. In this way, feudalism was a system of corresponding rights and responsibilities (kind of like *Confucianism* when you think about it). Finally, below the vassals were the peasants and **serfs**, and there were a whole lot of them. These were the common people who actually worked the land, but they didn't own it.

It's important to realize that no countries or government system existed in the way that we currently think about those things. Instead, the land was just sort of hanging out there, with kings claiming power over certain territories. People in one kingdom had characteristics in common, and so they were often intensely loyal to each other and to their culture just like in a modern nation or country. But people who lived in a kingdom did not really see themselves as countrymen. Instead, they identified with each other based on their social and economic class.

Peasants were directly responsible to their lords, who were in charge of the farms and farming villages, which were known as manors or **fiefs**. The peasants worked the land on behalf of the lord, and the lord gave them a place to live and protection in case of an invasion. Many of the manors were remarkably self-sufficient. Everything that was needed was produced right on the spot. Food was harvested, clothing and shoes were made, and so on. The lord, of course, was responsible to the king, but only when the king called upon the lord to provide a service. At all other times, the lord was the boss of his own manor.

When a lord died, his land usually was inherited based on **primogeniture**, which means the first-born son of the lord became the new lord. You would think, of course, that if a lord was killed, other surrounding lords might fight each other for control of the newly available manor, but that's not how it worked. The reason is that most of the lords (and knights, who were warriors that were also considered part of the nobility) followed the **code of chivalry**, which was an honor system that strongly condemned betrayal.

Feudalism existed for centuries in medieval Europe, but it slowly began to break down as agricultural production increased. Just as in ancient civilizations, in which cities grew once farmers were able to produce more food than they could consume by themselves, feudal manors grew when peasants were able to produce an excess of food, which freed other peasants to pursue other kinds of jobs (at the discretion of the lord, of course). Eventually, skilled craftsmen became commonplace, and then some of them started moving to towns.

As Europe started trading more frequently with the rest of the world, the skilled craftsmen started to earn extra income. When banking began in Europe, towns and cities really started to gain momentum. A "middle class" of craftsman and merchants emerged, which in turn lured more people into the towns to make money and develop a skill. The rise of towns, of course, upset the balance and stability of the feudal system. No longer were there clear lines of rights and responsibilities. No longer were all the peasants doing the same thing, and some of the peasants weren't even peasants any longer. They had worked their way up into a middle class.

> The changing social system set the stage for a completely different kind of Europe.

The big kingdoms

In eleventh-century Europe, England and France were the two major kingdoms, but they weren't countries in the way we think about them today. Instead, they were *kingdoms*, and because they were kingdoms, their existence was dependent on the loyalty of the nobility within their territories. The Holy Roman Empire encompassed a significant chunk of Europe, and included many German and Italian territories, but Germany and Italy weren't yet unified countries or kingdoms. There is more about all of this below in the section titled "The Emergence of Nation-States."

Rome out, Byzantine in

After the fall of Rome, Constantinople continued the traditions of organized Christianity as the capital of the Byzantine Empire (also known as the Eastern Roman Empire). But the Byzantine Empire was not a clone of the old Roman Empire. It used the Greek language; its architecture utilized distinctive domes; its culture had more in common with Eastern cultures like Persia; and its brand of Christianity, in which icons were an extremely important part of religious life, became an entirely separate branch known as Orthodox Christianity.

While much of Europe was fragmented into small feudal kingdoms with limited power and reduced intellectual activity (compared with what was going on at the height of the Roman Empire), the Byzantine Empire was one of the hottest things going. In virtually every aspect of cultural and political life, Byzantine peoples influenced the Middle East, eastern Europe, and Russia for hundreds of years, until Muslim Ottoman Turks captured Constantinople in 1453 and ended Christian dominance in the eastern Mediterranean area.

Meanwhile, down South . . .

During the Middle Ages, Islam was united in a single empire known as the **Islamic Empire**, kind of like the Byzantine Empire to the North, which united the territories under Orthodox Christianity. The Islamic Empire reached its height during the Abbasid Dynasty in the eighth through twelfth centuries. This was known as the golden age of Moslem (Muslim) culture. During this time, the Empire translated many ancient Greek and Roman texts, and expanded on mathematical and scientific knowledge gained from both the West and East. While knowledge in Europe was being lost during the Middle Ages, the Islamic Empire kept much of ancient Europe's past alive.

But the Islamic Empire was not only affected by other cultures, it affected those cultures as well. During this time, the **Moors** conquered much of Spain. With them, they brought knowledge of math, science, architecture, and literature to Europe.

> The lasting impact of the Islamic culture on Spain can be seen today in its architecture.

The playground is not big enough for two bullies

Meanwhile, up north, the Christians felt threatened by the expansion of the Muslims, especially as Islam became entrenched in areas that the Christians identified with historically. So, in 1096, Pope Urban initiated the **first crusade** in response to the success of the Seljuk Turks, who took control of the Holy Land (present-day Israel). The pope wanted Jerusalem, the most important city in Christianity, to be in the hands of Christians. He was also hoping that the efforts would help reunite the Roman Catholic Church with the Eastern Orthodox Church in Constantinople, which had split apart 50 years before the start of the Crusades. The crusaders immediately set out to conquer the Holy Land, and initially captured several cities, including Antioch and, most importantly, Jerusalem. However, both cities quickly fell back into the hands of the Arabs.

Through the year 1204, a total of four crusades failed to produce results, and the Eastern Orthodox Church and Roman Catholic Church separated even further (five more crusades also followed, but were not successful in achieving major goals). Most of the crusaders either died or returned to Europe. The impact on the Holy Land was a series of violence and uncertainty. Since most of the region remained in the hands of the Muslim Arabs, the series of crusades led to centuries of mistrust and intolerance between Christians and Muslims.

Quick Quiz 1

1 All of the following were characteristics of European feudalism EXCEPT

    A Peasants and lords had corresponding responsibilities.

    B Land was inherited based on primogeniture.

    C Cities were the main centers of activity.

    D Lords and knights followed the code of chivalry.

2 After the fall of Rome, organized Christianity was centered in

    F Constantinople

    G Jerusalem

    H Mecca

    J Antioch

3 During the Middle Ages, the Crusades were

Check your answers to Quick Quiz 1 on page 153. If you missed any of these questions, reread the previous section.

    A launched by Rome to defeat Constantinople

    B launched by European Christians to acquire land occupied by Muslims

    C launched by Muslims to convert Europeans to Islam

    D launched by Eastern Orthodox Christians to convert Roman Catholics

Conflicts Among Eurasian Powers

Conflict between European and Asian powers didn't stop with the Crusades. Two other big fights dominated the region at the close of the Middle Ages.

The mega Mongol Empire

At its peak, the Mongol Empire stretched from China all the way to Moscow. Regarded by Europeans as *barbarian*, Mongol culture did not enhance the arts, science, or the economy of the regions it conquered, such as Russia. This is not to say, however, that it did not affect the culture of Russia tremendously. It did. It brought Russian culture to a standstill. Individuality was not recognized. Instead, loyalty to authority was expected. Basic subsistence farming was practiced, and the general economic and social progress of Russia was slowed, even while the rest of Europe was emerging from the Middle Ages. This meant that Russia was behind the cultural, social, political, and technological developments that were beginning to re-emerge in Europe at the close of the Middle Ages. Russia has been trying to catch up to the West's economic, social, and political reforms ever since.

Here come the Turks

In the 1300s, Ottoman Turks invaded Asia Minor and were converted to Islam. They grew in power and, in 1453, they conquered Constantinople and changed its name to Istanbul. The city became the new seat of the **Ottoman Empire**, thus marking the downfall of the Christian Byzantine Empire and the rise of Islam in the region.

Ottoman rulers, known as *sultans*, were very powerful and organized. The most famous was **Suleiman the Magnificent** (1521–1566). He was known as *the Lawgiver* because he instituted a lot of legal and social changes. By 1566, the Ottoman Empire controlled most of the Middle East and part of Europe. The Ottomans tried to expand even deeper into Europe, but they were stopped at the **Battle of Vienna** in 1683.

The Emergence of Nation-States

Recall that during the Middle Ages, western Europe wasn't organized into countries (nation-states). Instead, it was broken up into feudal kingdoms. At the close of the Middle Ages, western Europe started to organize itself along cultural and linguistic lines. People who spoke Spanish viewed themselves as part of Spain. Those who spoke English united under the banner of England. You get the idea.

England emerges

As the first document to limit the power of a monarchy, the **Magna Carta** served as the first step toward modern representative governments and constitutional rights, and with it England emerged as a country built on more than just the power of a king. It was signed by **King John of England** in 1215, though he didn't sign it willingly. For years, he had ruled as a ruthless and self-serving tyrant, raising taxes on the nobility while restricting their privileges. The nobility organized against him. Endangered, King John fled to Runnymede, where the nobility caught up with him and forced him to sign the Magna Carta. Although it did not establish representative government or guarantee rights to the peasants, the document revolutionized the monarchy by limiting it. The document established rule of law and the beginnings of due process. Eventually, the English developed ideas from the Magna Carta and established a Bill of Rights and a representative parliament.

> The Magna Carta has served as the basis for the U.S. Constitution and hundreds of other documents that recognize inalienable rights.

France emerges

In 987, King Hugh Capet ruled only a small area around Paris. Over the next couple of hundred years, French kings expanded the territory, but beginning in the twelfth century, England began to gain large parts of present-day France. The English occupation of French-speaking territories led to French nationalism and a sense that France should be united under its own leadership.

At the age of 13, **Joan of Arc** heard voices that told her to liberate France from the hands of the English, who had literally claimed the entire French throne by the early 1400s. She convinced French authorities that she was divinely inspired to lead men in battle. With her army, she forced the British to retreat from Orleans, but was later captured, put on trial by the English, and burned at the stake. Nevertheless, she had a significant impact on the **Hundred Years' War** between England and France, which eventually resulted in England's withdrawal from France.

After the Hundred Years' War, royal power in France became more centralized. Under a series of monarchs known as the Bourbons, France unified and became a major power on the European continent.

Spain emerges

In the mid-fifteenth century, **Queen Isabella**, the ruler of Castille (present-day central Spain) faced a big problem. Power in the Spanish-speaking region of Europe was divided. There were two causes. First, Castille was one of three independent Spanish kingdoms, and therefore, no single ruler controlled the region. Second, the peasants were split along religious lines (mostly Christian and Muslim) due to remnants of the Muslim conquest of the Iberian Peninsula during the Middle Ages. To solve both problems, Isabella married **Ferdinand**, heir to the Spanish kingdom of Aragon, in 1469, thus uniting most of Spain under a single monarchy. Then, Isabella and Ferdinand, both Christians, made the Catholic Church a strong ally, as opposed to competing with the Church for authority. This alliance increased Spanish nationalism under the new monarchy by tying it to Christianity, thus ending religious toleration in the region. The result was that Muslims and the Jewish people were forced to convert to Christianity or else leave. This ushered in an era known as the **Spanish Inquisition**. The consequences for non-Christian Spaniards were tragic; the consequences for the Spanish monarchy were huge. Newly unified and energized, Spain embarked on an imperial quest that brought it tremendous wealth and glory, eventually resulting in the spread of the Spanish language, Spanish customs, and Christianity to much of the New World.

Russia emerges

You may recall that during the late Middle Ages, the Mongols conquered most of present-day Russia and stunted Russia's cultural and political development. By the fourteenth century, the Mongol power started to decline and the Russian princes of Muscovy grew in power. Muscovy was centered around the city of Moscow, which was the seat of the Russian Orthodox Church. The princes simply copied the ruling style of the Mongols and drove them out of Moscow, establishing a new feudal state. Nationalism in Russia, therefore, was fully underway.

By the late 1400s, Ivan III expanded Muscovy territory into much of modern-day Russia and declared himself *czar*, the Russian word for *emperor* or *Caesar*. As the center of the Eastern Orthodox Church, Moscow was declared the "Third Rome," after the real Rome and Constantinople (the "Second Rome"). By the mid-1500s, **Ivan the Terrible** (of the House of Rurik) had centralized power over the entire Russian sphere, ruling ruthlessly and using the secret police against his own nobles.

> Don't forget to practice using your Process of Elimination techniques when you answer the practice questions in this chapter.

Quick Quiz 2

1 The Mongol Empire

   A established many universities and museums in Russia

   B quickly westernized Russia's industries

   C was led by Peter the Great

   D slowed intellectual and cultural development in Russia

2 The Magna Carta is a document that

   F created the modern nation-state of France

   G led to the Spanish Inquisition

   H represents the beginning of democracy in England

   J brought an end to the Hundred Years' War

3 Which of the following was a method used by Isabella and Ferdinand to unite Spain?

   A They declared war on the Catholic Church.

   B They increased the power given to feudal lords.

   C They allied themselves with the Catholic Church and forced everyone to convert.

   D They reduced Spain's involvement in international affairs.

Check your answers to Quick Quiz 2 on page 153. If you missed any of these questions, reread the previous section.

The Renaissance

Greek and Roman influences

You might remember that during the Middle Ages, while Europe lost its Greek and Roman roots, the Islamic Empire rose. It was this empire that preserved and expanded upon Greek and Roman accomplishments. In other words, during the Middle Ages, most innovations occurred outside of Europe, especially when it came to math and science. During the European Renaissance (which literally means "rebirth"); however, the Greek and Roman roots of Europe were slowly rediscovered by the Europeans, in large part because Europe began to trade extensively with the Middle East. As the traders came in contact with each other, their ideas and philosophies merged.

Italian city-states: Home is where the art is

The Italian States During the Renaissance

In the fourteenth century, Italian city-states such as Florence and Milan grew rich from trade with the Middle East. These city-states were kind of like the city-states of ancient Rome and

Athens. Powerful families controlled cities and a small band of land around the cities that together acted like a mini-country. The powerful families in these city-states were intent on creating major centers of commerce, education, and the arts. The **Medici family** of Florence, for example, used its wealth to expand the arts, making Florence one of the most artistically beautiful cities in the world.

Renaissance men

The Italian Renaissance is known for its artistic achievements. Artists like Michelangelo, who painted the Sistine Chapel, and Raphael, known for his paintings of the Madonna, added a humanism to painting that was lacking in medieval European art. The Italian Renaissance, however, is perhaps best summarized in the work of **Leonardo da Vinci**. As a painter, sculptor, architect, engineer, writer, and more, da Vinci exemplified the extent to which the Italian Renaissance affected many aspects of life. His contributions to science and engineering, such as a simple machine using cogwheels, were practical and beautiful. Science permeated his art and art permeated his science. The Italian Renaissance, therefore, not only made significant contributions to the advancement of art and the advancement of science, but also to the combination of the two.

> Leonardo da Vinci is considered the premiere renaissance person because he excelled in so many differrent fields.

Clash with the Church

Italy was very artistic and also very religious. For a while, the Church embraced the Renaissance. Church officials employed artists to decorate cathedrals and monasteries. But the Renaissance also resulted in the development of **secular humanism**, which put the needs and pleasures of humans ahead of concerns for salvation and the afterlife. So during the Renaissance, the power of the Church slowly began to decrease. This led to major clashes, not only during the Reformation, discussed below, but also during the Scientific Revolution, discussed later in the chapter.

The Reformation

Martin Luther: A monk on a mission

For centuries before the rise of Martin Luther, the Roman Catholic Church controlled all religious worship and expression in western Europe. With this much power in the hands of one group, it shouldn't be surprising that the Church was engaging in activities that raised the eyebrows of some monks and priests. But when the Church hierarchy began to sell indulgences, Martin Luther, an Augustinian friar in Germany, could no longer contain his anger. **Indulgences** were pieces of paper that the Church sold to raise funds for its ambitious construction projects. The Church claimed that the purchase of an indulgence would result in reduced time in purgatory for the purchaser.

In response to this practice and others, Luther nailed his **95 Theses** to a church door in Wittenberg, Germany in 1517, pointing out the theological flaws in the Church's practices. Luther's ideas spread quickly, especially with the aid of the printing press, which had recently been invented. By the time the Church was able to excommunicate him in 1520, northern Europe had turned against the Catholic Church and declared itself Protestant. Other Protestant movements gained steam under the direction of John Calvin, Huldrych Zwingli, and others. Soon, several Protestant denominations broke away from the Catholic Church. Europe was no longer unified under the Catholic Church, but the Catholic Church that remained became stronger and even more centralized.

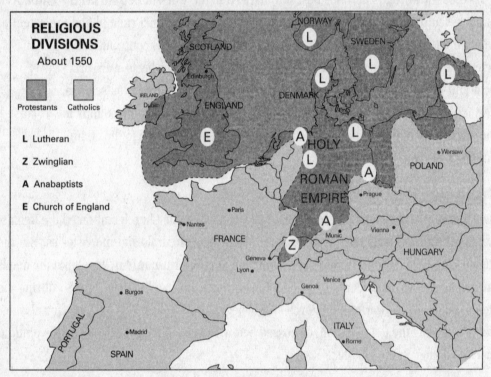

Religious Divisions Around 1550

John Calvin: Predestined to become a leader

John Calvin, a Frenchman who found a following in Geneva, also rejected many of the same Catholic principles as Martin Luther, but he had a profoundly different theology from Luther's. Calvin believed that the absolute depravity of man after the fall of Adam made salvation impossible except for the select few who were predestined by God. These few were known as *The Elect*. Only God, of course, knew who The Elect were, but Calvinists strove to live a holy life in the hope that they were included. In many senses, Calvin also changed Christianity by focusing on each individual's deeds and his or her relationship with God. He set up a theocratic state in Geneva, but his movement was hardly limited to Switzerland. It spread throughout much of Europe, challenging Catholicism and even Lutheranism. Major Calvinist movements won acceptance in Scotland (Presbyterians), Holland, England (Puritans), and parts of France (Huguenots).

John Calvin and Martin Luther, while enormously different from one another, challenged the authority of the Catholic Church and succeeded in establishing new denominations of Christianity. They refocused Christians in Europe on the words of the Bible, on the importance of their individual conduct not as an end in itself but as a sign of something else (either faith, for Lutherans, or as membership in The Elect, for Calvins), and on an individual's personal relationship with God without the aid of intermediaries.

> The movements that Martin Luther and John Calvin began still have millions of followers today.

Henry VIII: Divorce court

During the Middle Ages, England was Catholic, but as the Protestant Reformation got underway, Catholicism's hold on England began to weaken. The Catholic Church's frustration with England reached fever-pitch when Henry VIII was excommunicated from the Church after defying the authority of the Pope by divorcing his wife, Catherine of Aragon. Supported by many of the English people, Henry convinced the English Parliament to pass the **Act of Supremacy** in 1534, which created a national religion known as the Anglican Church of England, totally independent of the Pope.

Unity shattered and beliefs scattered

The Catholics responded to the Protestant Reformation with their own counter-reformation and were generally successful at reigning in their own troops. But the result was that Europe became divided along religious lines. Northern Europe and parts of western Europe were decidedly Protestant, while southern Europe and parts of central Europe were decidedly Catholic (parts of eastern Europe and Russia followed the Byzantine model of Eastern Orthodoxy).

Still, with all this division, some countries tried to find ways to accommodate all citizens, regardless of religious affiliation. In England, which in the late 1600s was controlled by Catholics, the **Toleration Act of 1689** granted some freedoms to various Protestant groups. In France, the **Edict of Nantes** in 1598 granted some freedoms to the French Huguenots, Protestants who had previously been discriminated against by the Catholic throne.

1 **During the European Renaissance**

   A Europe experienced a rebirth of Greek and Roman roots.

   B Russia emerged as the most powerful country in Europe.

   C Leonardo da Vinci led a reform movement within the Catholic Church.

   D The Medici family abandoned the arts and pushed Europe toward Protestantism.

2 **Who is associated with the 95 Theses?**

   F Martin Luther

   G John Calvin

   H Huldrych Zwingli

   J Henry VIII

3 **With the Act of Supremacy, England**

   A declared the supremacy of the Catholic Church

   B declared itself the supreme empire on the globe

   C began the Protestant Reformation

   D created the Anglican Church separate from the Catholic Church

Check your answers to Quick Quiz 3 on page 153. If you missed any of these questions, reread the previous section.

Age of Discovery: Cultures in the Americas, Africa, Asia, and the Middle East

Before you review the European discovery of the rest of the world, let's look at what the rest of the world was like before the Europeans arrived. There will be about eight questions about these parts of the world, both before and after the arrival of Europeans. In addition, there may be a few questions about major world religions and their influence on cultures prior to, as well as after, the European invasion. So, we've organized this section in the following way:

- The state of the rest of the world from about AD 1000 until the Europeans arrived

- European expansion into the rest of the world

- A quick review of major world religions

The State of the Rest of the World from about AD 1000 until the Europeans Arrived

India: Everything's nice when you add some spice

Around AD 1000, the golden days of India's Gupta Dynasty were over, but India was still a pretty interesting place to be. Waves of invasions by Huns were followed by waves of invasions by Muslims, mixing up the religious pot (part Hindu, part Muslim) and setting up the subcontinent for major religious division.

The invasion of the Muslims was momentous for India's economic development because India became part of the vast Muslim trade network. Spices began to flow from Southeast Asia and India back to the Middle East, and from there they were traded to Europe. When the Europeans realized that India and Southeast Asia were loaded with raw materials, they became interested in establishing direct trade ties with the region, rather than having to go through the Muslims. In the meantime, because of the increase in trade, Indian culture began to influence Southeast Asia. The **Khmer Empire**, a huge Hindu-influenced kingdom, was all the rage from about 800 through the mid-1400s in what is now Cambodia.

Around the 1500s, the **Mongols** invaded and established a vast empire centered in Delhi, where they ruled the country for the next 300 years. But by this time, India had been opened up to the rest of the world, and the Mongols were not able to fight off the advances of the world's other great powers. There is more information on this subject later in this chapter.

China: Show a little respect

In China around AD 1000, life was dominated by **Confucian philosophy**. Confucian philosophy centered around five key relationships: ruler and subject, father and son, husband and wife, elder brother and younger brother, and friend and friend. The idea was that within these relationships, each member owed the other member certain duties or attitudes. A father, for example, owed his son discipline, a safe place to live, and compassion. A son, in return, was expected to respect and obey his father. Key to this relationship was the idea of filial piety.

Filial piety, or behavior that is befitting a child who respects elders and legitimate authority, was a cornerstone of the social structure. Filial piety contributed to the maintenance of an orderly and harmonious society. Children were told that obedience was their primary goal, and that if they were obedient, they would grow up to be well-respected parents. Confucian society was replete with stories of obedient children, and these stories, of course, were repeated by elders. One story, for example, was of a dutiful boy who slept naked without covers to encourage the mosquitoes to feed on him rather than on his parents who were sleeping nearby. Such stories reminded children that it was their duty not only to listen to their parents and other legitimate authorities, but also to put the well-being of their parents and family above their own.

Of course, filial piety was meant to maintain order in society, but the tradition had many disadvantages as well. Some parents and authorities ruthlessly or carelessly put children in danger, but the children could not object because to do so would be to show disrespect. In addition, the process encouraged a lot of bad ideas to continue for longer than they might have had the society encouraged skepticism and competition rather than strict obedience. Therefore, filial piety sometimes led to unfortunate outcomes, especially when parents and authorities were not as true to their Confucian obligations as the children were to theirs.

Much attention was given to the responsibilities of those who ruled. Consider the **Mandate of Heaven**, which was similar to the Divine Right Theory that developed later in Europe, in which the ruler was considered to have a divine mandate to rule. Unlike the Divine Right Theory, the Mandate of Heaven did not justify absolute power and tyranny. Instead, the mandate was only good as long as the leaders ruled justly. As soon as they began to rule unjustly, they would be overthrown by a rebellion. If the rebellion was successful, it was determined that the leaders were no longer just and had "lost the mandate."

In terms of technological development, China was on the cutting edge in many ways. The Chinese produced silk, tea, porcelain, paper, block printing, gunpowder, and a mariner's compass long before anyone else. The rest of the world envied China's accomplishments. The world learned of China's advances via two methods: the **Silk Road**, a land-based route that ran from China through central Asia and into the Middle East, and the southern sea route, which ran from China, around India, and up to the Middle East.

> The Silk Road helped to spread cultural influence from the Middle East to Asia, and vise versa.

Japan: Shoguns show guns

Around AD 1000, the imperial period in Japan was on the decline and the feudal period was on the rise. This was because certain families in Japan were rising in power and weakening the central authority of the emperor. The emperor still existed, but he became much more of a ceremonial figure than a powerful one.

What's important about feudalism in Japan is that the power was divided into various classes, with the military warlords among the most powerful. The **shogun**, who was appointed by the emperor, was the military leader and the most powerful person in Japan. The **daimyo** was the landowner class, which was loyal to the shogun but also powerful in itself. The **samurai** was the warrior class, which supported the shogun and the daimyo militarily in return for land and supplies (sounds like European feudalism, doesn't it?). Peasants and artisans were below the samurai, and exchanged services for protection. Finally, the merchants were even lower on the social scale, though they did make a lot of money.

Interestingly, Japanese warriors were elevated in social distinction, while in China, for example, the scholars were elevated. This had a profound impact on social development in Japan. Unlike the Chinese, the Japanese were able to fight off invasions, which not only kept them separate and untouched while much of Asia fell to invasions but also fostered a militaristic attitude that would shape future generations. Of course, Japan's geography helped fight off invasions as well.

The samurai followed a **Code of Bushido**, which stressed simplicity, loyalty to one's lord, courage, and indifference to death. In fact, to avoid surrender or capture, warriors were expected to commit suicide.

Sub-Saharan Africa: Caravans through the sand

Around AD 1000, long before the Europeans arrived in sub-Sarharan Africa, a series of empires controlled most of the trade routes and intellectual activity.

During the fourteenth century, the **Mali Empire** in western Africa was at its height. Traders from Mali traveled throughout northern Africa on their way to the Middle East. The original leader of Mali—**Sundiata**—converted to the Islamic faith and all of Mali joined him. Since they traveled so extensively throughout northern Africa, the Mali introduced Islam to much of the region, converting small kingdoms as they passed through. The Emporer Musa even visited Mecca and Cairo, taking with him caravans of gold. The Mali brought back ideas and knowledge from the Middle East and Egypt. They also established **Timbuktu** as a major cultural destination, where people could learn not just about Mali culture but also about all the cultures with which the Mali had contact. But it was the cultural diffusion of the Islamic faith that probably had the most lasting impact on the region.

After the Mali were weakened by civil war, they were replaced by the **Songhai**, a civilization that also was heavily involved in trade and therefore made contact with other civilizations in Africa and the Middle East. During the Songhai's golden age, they accumulated and spread knowledge throughout the trade region. They traded gold to the Middle East in return for visits from Islamic scholars. Timbuktu remained a center of high culture and learning, complete with schools and libraries. Money from trade supported the construction of mosques and public buildings and the creation of art, and contributed to a prosperity unmarred by crime or significant violence.

Central and South America: Rigid social structures

While the Mayans built the most significant empire in the Central America before AD 1000, the post-AD 1000 Americas were dominated by the Aztecs and Incas.

The **Aztecs** were warriors who were big on rigid social structure. They were skilled engineers and architects. They forced conquered peoples to pay tribute to prevent war (paying tribute included supplying slaves for sacrifice). **Tenochtitlán**, the Aztec capital, was close to present-day Mexico City. The Aztecs were ruled by a warrior king, who was supported by a warrior class and a priestly class. Unlike the Europeans, the Aztecs practiced polytheism. Particularly important was the god of war, Huitzilopochtli, who could only be worshipped properly by human sacrifice. In short, the Aztecs were the center of Central American civilization before the Europeans arrived.

The **Incas** were located in the Andes of South America, stretching from southern Colombia through Ecuador, Bolivia, Peru, northern Chile, and northwestern Argentina. They numbered in the several millions. The Incan emperor, who was revered as a descendent of the sun god, controlled everything through a network of officials and priests. The Incas were road-builders, engineers, and even surgeons. They also had extensive farms and food distribution systems.

> Like the Muslims, the Incas placed great emphasis on knowledge and scholarship.

European Expansion into the Rest of the World

Explorers and the conquistadors: Our way or the highway

Both Spain and Portugal financed explorations of the open seas to find the shortest possible path to Asia, where they hoped to gain access to Asian resources and products. Spain sponsored the voyage of **Christopher Columbus**, for example, in an attempt to find a route by sailing west across the Atlantic, but of course, Columbus ended up in Latin America instead. When it was later realized that Columbus had, in fact, not arrived in Asia like he thought he had, both Spain and Portugal funded expeditions to explore the resources of the vast and unknown continent.

The Spanish, who had grand colonization plans for the region, did not merely explore, but also engaged in military conquest. When the Spanish began to attack, they wiped out the Aztecs and the Incas. European diseases, like smallpox, were introduced to the native population resulting in major losses of life. Furthermore, the Spaniard's superior military weapons and use of horses enabled them to inflict devastating assaults on the Aztec and Inca Empires. At first, the Aztecs didn't even fight the Spanish, since they believed that **Hernando Cortes**, the Spanish conquistador who first made contact with them, might be a god. This allowed the Spanish to build up their forces with little opposition. The Spaniards used other native groups that they had previously conquered to help them annihilate the vast Aztec Empire. Spain became the dominant empire on the continent, and Portugal took control of present-day Brazil. Eventually, the entire region was altered drastically by the influx of European culture, religion, politics, and art, and Spain became the dominant power in Europe until the rise of Great Britain.

Social classes

As the Spaniards colonized Latin America, they exploited the tremendous agricultural and mining potential of the region. These opportunities for wealth were some of the main reasons for the colonization effort in the first place. Not wanting to do the hard labor themselves, and yet needing a large number of workers to complete the tasks of farming and mining, the Spaniards devised a system known as *encomienda*, under which they forced the natives to do the work. Most of the Spanish landlords treated the natives as slaves. As a result of the rigors of the system and the spread of European diseases within the exhausted slave community, the Native American population dropped sharply.

European expansion into Africa

Faced with a shortage of labor in the Americas, European and American colonists turned to Africa for replacement labor. The Africans were captured on the continent's interior and traded like commodities to the Americas. African slaves were in great demand because the plantation systems required thousands of hard laborers. The impact of the slave trade on the Americas was enormous, especially in parts of Central America, where African cultural traditions took root. As a result, the Americas became more ethnically diverse. As for Africa, of course the impact was enormous as well. Not only was Africa drained of millions of citizens, but contact between Africa and Europe was fully underway. Later, during the Industrial Revolution, Europe would become more interested in the continent and actually colonize it for itself.

European expansion into Asia and the Middle East

The colonization of the Americas didn't satisfy the Europeans' desires, however. They still eyed Asia closely. Trade efforts were constantly in the works, but it wasn't until the Industrial Revolution that Europe was bold enough to actually demand access to the continent. Expansion into Asia will be discussed later in the section on the Industrial Revolution.

> Throughout history, trade has always affected culture, and vice versa.

Influence of Christianity

During colonization of the Americas, Christianity played an enormous role. First, since the Europeans were Christian, they brought their religion with them to the New World, and therefore altered the religious and social landscape of the New World forever. Second, since Christianity is generally an evangelical religion (which means that its adherents believe that it should be spread), many Christians considered their religious beliefs a reason to colonize in the first place. They worked to convert non-Christians to their faith. Many Europeans were motivated by economics and power, but since Christianity fit nicely with their goals, they embraced it and won the approval of the Church, and with it millions of loyal Church adherents. Christianity also played an enormous role in later European colonization efforts in Africa and Asia during the Industrial Revolution. More on that later.

Quick Quiz 4

1 **All of the following groups invaded India from AD 1000 to 1600 EXCEPT**

   A Huns

   B Muslims

   C British

   D Mongols

2 **Filial piety means**

   F showing respect to Buddha

   G showing respect to parents and other legitimate authorities

   H showing respect to conquistadors

   J showing respect to the shogun

3 **Which list puts the feudal Japanese classes in the proper order from highest to lowest?**

   A shogun, daimyo/samurai, artisans, merchants

   B shogun, artisans, merchants, daimyo/samurai

   C daimyo/samurai, shogun, merchants, artisans

   D daimyo/samurai, shogun, artisans, merchants

4 **Which of the following was a reason that the Europeans went to Africa in search of slave labor?**

   F It was cheaper to use African slaves than natives from the Americas.

   G Disease killed many of the natives in the Americas.

   H Europeans were over-worked in the Americas.

   J Europeans had heard that Africans were very strong.

Check your answers to Quick Quiz 4 on page 153. If you missed any of these questions, reread the previous section.

A Quick Review of Major World Religions

Throughout the Middle Ages and afterward, the world's major religions took root. They became intertwined with political governments; influenced political, military, economic, and artistic goals; and won the allegiance of hundreds of millions of people, who in many cases placed their religious beliefs above all else.

Since religion during the second millennium AD was so often the cause of controversy and conflict, the test writers will likely ask you a question or two about it. You need to understand both the basic tenants of the major religions and a bit about the influence these religions have had on the development of civilizations. The test writers will focus on Judaism, Christianity, Islam, Hinduism, and Buddhism. Questions about the effects of these religions may cover any period from medieval times to current times, so we've included a quick discussion of the impact of religion.

> As you read the sections that follow, try to keep in mind the impact of religion on history.

Judaism: A religion and an ethnicity

Judaism is an ancient religion of the Middle East, specifically Palestine. The Jewish people are monotheistic and believe that life should be spent living the will of God, as told to Moses. There are many subgroups within Judaism, as within all major religions, but the belief that they are the chosen people of God is central to the faith. It was this belief that unified the Jewish people even when they were dispersed throughout the world during the **Diaspora**.

Throughout the Middle Ages and since, the Jewish people have been persecuted and separated primarily from Christian groups in Europe, North America, and North Africa. But many never lost touch with their faith. Many believe that persecution and atrocities, such as the Holocaust, strengthened the faith of the Jewish people and convinced them that they should have a homeland of their own where they could be reunited. The **Zionist movement**, which was built on the idea that the Jewish people are chosen by God, advocated the creation of a Jewish homeland in Palestine, which is the land that the Jewish people believe was promised to Moses by God. As a result, the modern nation of Israel was established in 1948 as a Jewish state. But Israel was and is surrounded by Muslim nations. The creation of the state of Israel has led to persistent violence. There is more information about this later in this chapter.

> The next time you watch the news or pick up a newspaper and hear about troubles in the Middle East, remember how many years these troubles have been going on.

Christianity: Spread far and wide

Christians believe that Jesus Christ is the Son of God and that forgiveness of sins, and ultimately everlasting life, can be achieved only through belief in the divinity, death, and resurrection of Christ. Many Christians also believe that it is their duty to spread this message to non-Christians.

After Constantine converted to Christianity, the Christian hierarchy dominated not only the religious activity in Europe but also the political and social activity. Christians were not tolerant of other religions and so embarked on Crusades to convert the Muslims who had taken control of the Holy Land. Ultimately, the Crusades were not successful. Yet, even after the Reformation, both Protestants and Catholics (the two main divisions of western Christianity) remained committed to the conversion of others. This fit in nicely with the imperialistic plans of the European monarchies. The monarchies gained overwhelming public support for their invasions of established nations in the Americas, Africa, and Asia by justifying the actions as a way to convert heathens to Christianity. Many Europeans supported imperialism as a moral and religious duty, in some cases advocating any means available to accomplish their end.

Many Christians, of course, practiced their religion faithfully and consequently converted others to Christianity based on the strength of their faith and the merits of their religion. Nevertheless, imperialists and mercantilists were not above using the religion of the people to support their political and economic agendas. And so as Europe colonized the world, it took Christianity with it, and Christianity took root in the Americas, Africa, and, to a lesser degree, Asia.

Islam: Spread far and wide, part II

Islam means "submission," or in the religious context, "submission to the will of God." Muslims believe that Muhammad, their founder, is God's prophet. This belief is best expressed in the first of the **Five Pillars of Islam**: "There is but one God and Muhammad is his messenger." The other Pillars define proper Islamic conduct, as does the **Koran**(or Qur´an), the most important text of the religion. The Koran is extremely important since Muslims are focused on submitting to the will of God, and the Koran defines that will. As a result, the religion has tended to nurture fundamentalism. In other words, many Muslims try to follow the teachings of the Koran as closely as possible.

This fact has greatly influenced culture in the Middle East, where Islam is the dominant religion. Not only have many Islamic people resisted Westernization because it is incompatible with the teachings of the Koran, but they have created state religions so that governments follow and enforce the teachings of the Koran as well. Most notably, Iran reversed its westernization attempts after the fundamentalist **Islamic Revolution of 1979** made the Koran the governing text of the country. Throughout the Middle East, Islamic fundamentalism is on the rise because many Muslims do not want societal changes to result in their failure to practice what they believe to be the will of God. There is more information on this subject later in the chapter.

> History shows that religion divides peoples as often and as passionately as it brings people together.

Hinduism: Rewards for good behavior

Hinduism is a traditional religion in the strictest sense. In other words, it has been passed from generation to generation, not through written text or through a central religious authority, but as a matter of custom and tradition. Hindus believe that the person you are in this life was determined by the person you were in a past life. In addition, Hindus believe that how you conduct yourself in your assigned role in this life will later determine the role (called a *caste*) you get in a later life. This belief has affected the social, political, and economic agenda of India enormously.

The **caste system** in India is based on the Hindu religion. It was outlawed when India became an independent nation, but millions still practice it as part of their religion. The people of the lowest caste (called the *untouchables*) cannot mingle socially or even work with members of the higher castes. This has led not only to widespread social discrimination but also to employment discrimination. Many Hindus do not challenge discrimination because they want to be rewarded with a higher caste in a later life. Social mobility is greatly hindered. As more people move to the cities, they begin to disassociate themselves from the caste system. However, Hinduism remains a tremendous influence in the lives of millions of Indians.

Buddhism: Detach and be happy

Central to Buddhism are the **Four Noble Truths**: Life in this world involves unhappiness; unhappiness is caused by the desire for worldly things; happiness can be achieved by detaching oneself from these worldly things; **nirvana** can be reached for those who follow the **Eightfold Path** (rules of conduct and thought). This system of thought has greatly effected the development of Southeast Asia. For centuries, the region remained relatively isolationist, content with subsistence farming methods and limited material wealth. Its wars were fought for control over land and people, but not for control of natural resources or industry. The region was dragged into the Industrial Revolution by European imperialists, but not before putting up a fight. Even today, as the region's economies rapidly expand, the typical person lives without attachment to material things, unlike many of his or her Western counterparts.

Quick Quiz 5

1   **Which religion is most closely associated with traditional India and the caste system?**

   A   Buddhism

   B   Christianity

   C   Hinduism

   D   Judaism

2   **Which religion is most closely associated with the Zionist movement?**

   F   Buddhism

   G   Christianity

   H   Islam

   J   Judaism

3   **Which religion is most closely associated with European imperialism?**

   A   Buddhism

   B   Christianity

   C   Islam

   D   Judaism

4   **Which religion is most closely associated with the Four Noble Truths?**

   F   Buddhism

   G   Christianity

   H   Islam

   J   Hinduism

Check your answers to Quick Quiz 5 on page 153. If you missed any of these questions, reread the previous section.

Sixteenth Through Nineteenth Centuries: Enlightenment, Absolutism, Reason, and the Industrial Revolution

The sixteenth through nineteenth centuries were jam-packed with political, cultural, intellectual, and economic developments. We'll focus on the concepts and events that the test writers will most likely stress. Keep in mind that much of this section is devoted to Europe since the test writers will stress European developments more than others. Here's what is covered:

- Europe in the sixteenth, seventeenth, and eighteenth centuries

- Nineteenth-century political developments

- The Industrial Revolution

There will be 12 questions from the material covered in this section, so know it well.

Europe in the Sixteenth, Seventeenth, and Eighteenth Centuries

Absolute monarchs: Power hungry to the hilt

The power in an absolute monarchy rests with one individual, the monarch, who earns the right to govern by birth and who rules "absolutely." Many absolute monarchies justified their power to rule under the **Divine Right Theory**, which asserts that the monarchy is an instrument of God's will (like the Mandate of Heaven in China). Argue with the monarchy, so the theory goes, and you argue with God. One of the benefits of absolute monarchies is that they are generally very stable. If the nation is blessed with a string of competent monarchs, long-range plans can be carried out without being watered down by a large governmental body that tries to accomplish too much at once, or worse, tries to accomplish inconsistent goals. Since authority is not questioned and responsibilities do not overlap, the government can run very smoothly as long as everyone does his or her job.

Of course, absolute monarchies can have an extremely negative impact on the development of a nation if the nation is cursed with even one incompetent or ruthless monarch. Since the monarch has absolute authority, human rights can be violated wildly and the long-term interests of the nation can be irreparably harmed. Therefore, the success or failure of an absolute monarchy primarily depends on the personality and competence of the individual monarch.

Absolute monarchies were common in Europe before the Enlightenment as well as in China throughout much of its history. In France, the best example of a monarch claiming to rule by divine right was **Louis XIV**, who believed himself to be so much at the center of his kingdom that he became known as the "Sun King." Just as the earth depends on the sun, he argued, so the kingdom of France depended of the Sun King.

Louis XIV's reign under divine right had a tremendous impact on France. Louis centralized political and military power under his own control, taking substantial power away from feudal lords. All French soldiers, who previously fought for their feudal lords, now fought directly for the king. In addition, local parliaments were stripped of their autonomy, expected to simply rubber-stamp all of the king's laws. If they refused, the king would personally appear to enforce his decisions. Divine right, therefore, led to the centralization of authority under the monarch. This centralization led to excesses of power, which 100 years later, the masses would no longer tolerate. They revolted against absolute rule in one of the most significant events in Western history, the French Revolution.

In Britain, **Elizabeth I** secured Britain's position as a world power by ruling absolutely. Britain had rights established by the Magna Carta and the Parliament, but Elizabeth exercised great power when she needed to. She sponsored explorations and increased trade with other nations. When Spain threatened to invade and establish England as a Catholic nation once and for all, she defeated the **Spanish Armada** in 1588, thereby securing Britain as the world's greatest naval power. Long term, Elizabeth's reign set the stage for British imperialism. Under her leadership, Britain developed the military and commercial capacity that was used later to colonize much of the world.

> Like religious leaders, political leaders of this era believed they had the right to rule thanks to God.

The Glorious Revolution: Absolutism fades

When King James of England wanted to establish an absolute monarchy to eliminate the need for a parliament and reassert the Roman Catholic Church as the state religion, the Parliament supported the forces of **William of the Netherlands**, who'd fought against the Catholic Irish forces advocated by King James. James eventually fled and his daughter, Mary, took over the throne with her new husband, William. The event is known as the Glorious Revolution because it resulted in the formation of a limited monarchy with considerable power granted to the Parliament and increased rights for the general citizenry.

The long-term impact was even more considerable. It began a European movement against absolute monarchies and increased representation by the people. In addition, the **English Bill of Rights**, which was signed by the new monarchy and Parliament in 1689, served as a basis for the United States Bill of Rights and the wave of constitutional limitations on power that has swept the world ever since.

Enlightenment writers: Power to the people

John Locke believed that the natural rights of humans are best protected in representative governments. He wrote that the power of government should be limited by the will of the people. Therefore, if the people are dissatisfied with the decisions and policies of their leaders, they should be able to elect new ones. This philosophy represented a change in political thought. Before Locke and the other Enlightenment thinkers, monarchies claimed that their power came directly from God and that the purpose of the citizenry, therefore, was to accomplish the will of God as expressed through the monarchy. Locke challenged this justification for power and argued that the government is granted its right to govern from the governed. As such, he argued, it is the will of the people that the government must respect. In doing so, however, not even the majority of people can act through the government to deny an individual his basic rights to life, liberty, and property. This type of government, Locke argued, is not only morally right, but leads to a society of people whose individuality can be expressed and whose potential can be realized. His writings inspired the Declaration of Independence and Constitution of the United States as well as the constitutions of many other republics.

In his book *The Spirit of the Laws*, **Baron de Montesquieu** argued for separation of powers within government in order to prevent absolute monarchs like Louis XIV from seizing control of an entire nation of people. In concert with the works of other Enlightenment writers, his writings greatly influenced the popular opinion regarding the best form of government, thereby weakening the arguments of monarchs who claimed they had a divine right to absolute rule. In the centuries since, constitutional monarchies and democracies, both of which separate powers within the government, have become the most common forms of government in the world.

The French Revolution: European monarchs freak out

The storming of the Bastille, which set the French Revolution in motion, occurred because of middle and lower class anger with the French nobility. The middle class consisted of merchants and professionals who were experiencing an increase in wealth but also an increase in taxes. Together with the peasant class, who were not experiencing an increase in wealth but who *were* paying excessive taxes, the middle class wanted more political power in the **Estates General**, a parliamentary body comprised mostly of the nobility. They were motivated by the ideas of representative government and the rights of humans expressed by the Enlightenment writers and by the events in the American Revolution. Basically, they wanted one-person, one-vote. When they didn't get it, a riot broke out and a group of peasants stormed the Bastille, a political prison where enemies of the monarchy were kept, and let the prisoners out. One the French Revolution started, rioting quickly swept the nation. The storming of the Bastille did not solve any of the problems of the peasants and middle class, but it did set the revolution in motion.

The French Revolution began as an anti-monarchy campaign, but it didn't stay that way for long. It soon took on a nationalistic fervor when other European monarchies freaked out. They were fearful that if the peasants in France could revolt, their own peasants could revolt. So the other European monarchs threatened invasions against the French to try to get them to stop their popular uprising. But it didn't work. The French were united by nationalism. By 1793, the **Jacobins** (French revolutionaries who wanted to make radical changes) had murdered the members of their monarchy, including Louis XVI, and successfully prevented invasions from neighboring monarchies.

Then things got really weird. The Jacobins seized full power in the country and established the **Committee of Public Safety**. This group was very aggressive and very committed to its cause. They were so committed to establishing a country where the Rights of Man were respected that, in the process, they managed to trample over the rights of everyone in sight, including their own people. It was blind patriotism. The committee, under the leadership of **Maximilien de Robespierre**, began a **reign of terror** in which those suspected of opposing the revolution were executed. Kind of creepy, huh? The victims had turned into the torturers. The Jacobins even instituted the first military draft in European history to fight off the invasions coming from their freaked-out neighbors.

Moderates within the French rank-and-file began to realize that the Committee of Public Safety was going way too far, and so they united and turned on the Jacobins. They sent Robespierre and friends to the guillotine in 1795. They wrote a new constitution and established France as a republic under the leadership of a five-member Directory. The Directory wasn't very successful and was beaten up by Russia and Austria in war; so it was overthrown by an ambitious general named **Napoleon Bonaparte**, who claimed to be the true son of the revolution. He set up a military dictatorship (he declared himself "Emperor of France") and brought an end to the revolution.

> If it weren't for the chaos of the French Revolution, Napoleon would probably never have risen to power so quickly and absolutely.

The impact of Napoleon: Big things in small packages

Napoleon reigned in France from 1799 to 1815. Although he was a dictator, he instituted a lot of reforms that had a lasting impact on the development of democracy in Europe. His centralized government, called the Consultate, did in fact restrict personal freedom, but it promoted equality under the law. The **Napoleonic Code**, laws that codified some of the basic tenants of the revolution—such as trial by jury and religious freedom—changed legal documents forever. He also abolished discrimination against the Jewish people, created the Bank of France, and set up ambitious public schools.

But it was his hunger for military conquest that would eventually bring him down. He stormed though much of central Europe and claimed tons of land in the name of France, but once he reached Russia, his army was severely weakened. He did not take Russia, and an alliance led by England's Duke of Wellington defeated him at the **Battle of Waterloo** near Brussels, Belgium, in 1815.

Nevertheless, by this time, the ideas of the French Revolution had been carried to the rest of Europe by Napoleon. These ideas included constitutional government, rule of law, abolishment of the feudal system, and perhaps most importantly, nationalism. Long term, these ideas influenced the development of stable democracies in much of western Europe long after Napoleon was defeated. Although he was a dictator, Napoleon's civil code and educational reforms were copied in much of Europe as ways of maintaining order while increasing opportunities.

Scientific revolution: Prove it or lose it

Nicolas Copernicus was a sixteenth-century Polish astronomer who believed that the Sun, not the Earth, was at the center of the known universe. This view was in sharp contrast to the **Ptolemaic model** that had dominated the Western world's understanding of the universe. Ptolemy, an ancient Greek astronomer, believed that the Earth was at the center of the universe. Science and philosophy had developed for 1,500 years under this assumption, as had the Catholic Church for 13 centuries. The Church challenged Copernicus, claiming his theory reduced the traditional role of man as the center of God's creation. The Catholic Church insisted that since man was made in the image of God, the Sun was made to serve the interests of man, and this meant that the Sun must revolve around the Earth. Nevertheless, Copernicus continued to study the solar system. Those who followed him, including Galileo, kept the pressure on the Church and the academic community to rethink their worldview.

Using his telescope, **Galileo Galilei** observed the patterns of the stars and the relationship of the Sun to the Earth and the Moon to conclude that the Earth travels around the Sun. His telescope allowed him to get better evidence than even Copernicus had. Meanwhile, Pope Paul V defended the Church's position that the Earth is at the center of the universe and that the Sun travels around the earth. Galileo used scientific investigation to reach his conclusions whereas Pope Paul V used the position of the Church and faith to reach his conclusion. Therefore, they disagreed not only on the relationship between the Sun and the Earth, but also on the proper method for acquiring knowledge. Pope Paul V brought Galileo to Rome on charges of heresy, and Galileo was forced to recant his findings. But the debate over scientific investigation and the findings that resulted from it had by this time exploded onto the international scene, and there was no stopping the Scientific Revolution.

To understand the significance, you simply need to look at the development of the **compass** and **astrolabe**. The consequences of the development of these two mechanisms are so far-reaching that it seems safe to say that every nation in the world has been affected. In some cases nations owe their existence to these two devices. The astrolabe, which measures the altitude of celestial bodies, and the compass, which utilizes the Earth's magnetic field to give direction parallel to the Earth's surface, were essential navigational aids that made the Age of Exploration and the subsequent Age of Imperialism possible. Used by the navies of the European powers, these instruments led to the destruction of native civilizations in the Americas, Africa, Asia, and Australia by allowing hundreds of ships to make repeated voyages to the same destinations. This led to the indisputable military and commercial supremacy of Europe for centuries and changed the political and cultural landscape of everywhere else. Isolationism, therefore, became almost impossible after these two instruments were perfected.

Also during the fifteenth century, **Johan Gütenberg** invented movable type, and the printing press was born. This affected Europe in a profound way. Because of the printing press, the ideas of Luther's Reformation spread rapidly, leading to the reorganization of Europe along religious lines. The Bible was printed in popular languages such as German, allowing people to read and interpret it for themselves rather than relying on the Church hierarchy, thus contributing to individualism. This individualism led to the Enlightenment, which questioned the role of government authority and consequently changed the relationships between humans. All the while, literacy increased dramatically, books became less expensive, and education became an important goal for millions. In short, the printing press revolutionized Europe and the rest of the world because it brought information to the masses. If knowledge is power, the printing press resulted in a redistribution of power.

> You're about halfway done with your review of history from AD 1000 to the present. Congratulations. **Remember:** It will all be worth it when you score well on the SOL exam.

Developments in Russia: Romanov's greatest hits

After the fall of the House of Rurik in Russia, Czar Michael Romanov took over in 1613, establishing the **Romanov Dynasty** that would rule Russia for the next 300 years. One of the most influential Romanov rulers was **Peter the Great**, who was intent on expanding and westernizing Russia. He conquered the Baltic region from Sweden, where he built his "window to the West," St. Petersburg, as the new capital of Russia. He brought in Western scientists, architects, designers, and businessmen, learned about the West, and then transferred this new knowledge directly to the newly conquered region. The whole city of St. Petersburg was built specifically as a Western city! The impact for the Baltic region was enormous. Suddenly, it was at the center of activity in the Russian Empire. While the rest of Russia essentially dragged its feet in Peter the Great's westernization attempt, the Baltic region became the symbol of the new Russia.

Russia's westernization attempt didn't stop with Peter. **Catherine the Great**, empress of Russia for over 30 years in the late eighteenth century, aggressively used her authority to westernize and expand her empire. She attacked the declining Ottoman Empire to the south and secured access to the Black Sea for Russia. In addition, she successfully fought wars and negotiated agreements to win parts of Poland, the Ukraine, and Lithuania for the Russian Empire. Domestically, Catherine supported education and development of the arts, granted freedom of religion, but decentralized the power that Peter the Great had worked to amass, returning a considerable amount of power to the landlords. This kept Russia a little behind the times while western Europe's Scientific Revolution grew into the Industrial Revolution.

Quick Quiz 6

1 **Which of the following best describes Divine Right Theory?**

  A Monarchs believe their right to rule is given to them directly by God.

  B Monarchs believe their first-born child is entitled to the throne.

  C Monarchs believe they have the right to acquire neighboring empires.

  D Monarchs believe they should only rule as long as the people consent.

2 **The Glorious Revolution is significant to British history because it**

  F established Catholicism as the official state religion, thereby ending religious division

  G kept all power solidly in the hands of the monarch

  H freed political prisoners locked up in the Bastille

  J limited the power of the monarchy and established the English Bill of Rights

3 **The goal of the Enlightenment writers was to**

  A support the European monarchs

  B increase support for European military conquests

  C limit the power of the government to interfere with the rights of the people

  D uphold the Divine Right Theory

4 **Which of the following is true of the French Revolution?**

  F After overtaking the monarchy, the French immediately established a democracy.

  G Very few people were killed at the hands of the revolutionaries.

  H The French Revolution didn't achieve its own goals until after Napoleon's reign.

  J The Jacobins wanted to restore peace by slowly moving from a monarchy to a democracy.

5 **Which of the following was an accomplishment of Napoleon?**

  A the establishment of a code of laws

  B the conquest of Russia

  C the establishment of a democracy

  D the advancement of Europeanism over nationalism

6 Who challenged the Catholic Church's view of the Earth's position in the solar system?

F Galileo

G Luther

H Locke

J Robespierre

7 Peter the Great's impact on Russia is best expressed by the word

A feudalism

B westernization

C orthodoxy

D appeasement

Check your answers to Quick Quiz 6 on page 153. If you missed any of these questions, reread the previous section.

Nineteenth-Century Political Developments

The Congress of Vienna: Pencils and erasers at work

After Napoleon's defeat, England, Russia, Prussia, France, and Austria got together at the Congress of Vienna (1814–1815) to draw up peace plans and redraw the map of Europe. Led by Prince Klemens von Metternich of Austria, this group wanted to return things to the way they were before the French Revolution, complete with class distinctions and special privileges. Their goal was for no nation to be powerful enough to threaten the security of another. Of course, reaching this goal was quite complicated. Territory had to be shifted, and with territory shift came compensation. In other words, if the Congress took territory from a power, it often took territory from a different power to compensate. So the map of Europe was pretty much redrawn, with the big powers swapping land.

A whole lot of people didn't like what was going on. After all, the Congress of Vienna was an attempt to restore power to the elite and to bring order within the lower classes. Nationalistic revolutions swelled with the Germans, Poles, Italians, and Greeks, who wanted boundaries and nations to be based on ethnicity and nationalistic commonality, not on the political agendas of the leaders.

Unification of Italy: Italians give Austrians the boot

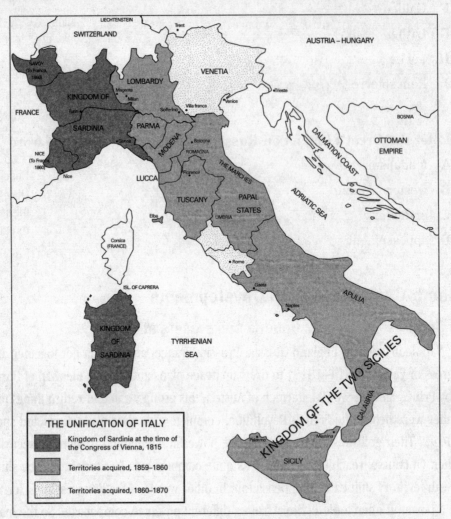

The Unification of Italy

Prior to the mid-1860s, the Italian peninsula wasn't a country, but rather a series of city-states. Some of these states were focused on commerce and art; others were focused on agriculture. The states of central Italy were papal, meaning they were owned by the Catholic Church. As nationalism swept Europe, especially at the time of Napoleon, the Italian peninsula began to unite under its common language and culture. Individuals like **Giuseppe Mazzini** formed an organization called **Young Italy**, which was dedicated to the removal of Austrian influence and control on the northern peninsula.

Later, in 1852, **Count Camillo Cavour** became prime minister of Sardinia-Piedmont, limiting the power of the monarchy in the small Italian kingdom and demonstrating to the other Italian states that democracy could work. Cavour then aroused nationalism on the entire Italian peninsula, telling them to "put aside all petty differences" and unite. Under an army organized by **Giuseppe Garibaldi**, the various Italian states drove out foreign influences and began to come together under a single Italian banner. By 1861, the kingdom of Italy was officially organized as a limited monarchy. By 1870, Rome became the capital city of the new country.

Unification of Germany: All about Otto

Otto von Bismarck united the independent German states into the single nation of Germany. In the nineteenth century, Germany was still a loose collection of city-states while much of Europe (France, Britain, and Spain, for example) had long been unified under a single government. Bismarck became the chief minister of Prussia, the largest German state, in 1862 and began his quest for unification. Many of the German states, however, were under the control of Austria. Thus, he first enlisted the support of Austria in a war against Denmark in order to gain control of two small German states. After succeeding, Bismarck turned against Austria (he didn't need them anymore) and defeated the country soundly, winning control of more German states.

The rest of Europe became nervous at the prospect of a united Germany, especially France. Bismarck then provoked France into declaring war on Prussia so that the rest of Europe wouldn't think that Prussia was aggressive. Bismarck then easily crushed the French in the Franco-Prussian War, and Germany unified, with Bismarck as its fearless leader. The result was not only the restructuring of the balance of power in Europe, but also the development of a strong sense of nationalism in Germany, which proved to be a very dangerous force just decades later in both world wars.

> At this time, the idea of nationalism was becoming move and more popular in Europe.

The Industrial Revolution

Western Europe was well suited for the emerging Industrial Revolution. Its natural harbors, supply of coal, and high population density allowed industrialism to spread quickly.

The Industrial Revolution is so important and had an impact on so many aspects of personal life as well as international politics that no summary can do it justice. Still, here we go . . .

Technological Developments: The Little Engine that Could

Two developments in the Industrial Revolution changed transportation and European economies, dramatically. The first was the invention of the **steam engine**. The second was the development of improved methods of smelting iron ore. Together, these two developments resulted in the invention of the railroad, and Europe and the rest of the world would never be the same. Raw materials were sent to hundreds of locations. Finished products radiated out of factories in all directions. But perhaps most significantly, people were suddenly on the move like never before. And with them, ideas and cultures spread, en masse, to new areas. The development of the railroad aided cultural diffusion more than any other invention up until that point in time (the invention of the radio, television, and the Internet probably have surpassed it).

The steam engine was used for more than transportation. It was also the catalyst of the factory system, generating enough power to run huge machines that humans operated in assembly-line fashion.

The effects of the factory system: Cough, choke, gag

The factory system brought previously unimaginable amounts of consumer goods to the western European population. Over time, this enhanced the quality of life for millions by providing inexpensive and easy access to a vast array of consumer products ranging from light bulbs and shoes to medicines and toys. But this occurred at a great social cost.

Prior to the Industrial Revolution, most European families worked on farms. They worked long hours, especially during the planting and harvesting seasons, but they worked together as families. Everyone knew what was expected of them—men, women and children. Although education was becoming increasingly available, the school schedule reflected farming seasons so children would be able to help their parents. The Industrial Revolution, however, sent millions of families to the cities where both parents had to work to make enough money to support the family, but unlike life on the farms, the parents often worked in different factories. Children often worked in factories as well, separating the family.

Whereas the farms provided exposure to clean air and sunshine, the factories exposed the workers to air pollution, hazardous materials, and machinery. Whereas the farms provided seasonal adjustments to the work pattern, the factories spit out the same finished products day after day, all year long. All of this eventually led to the development of unions and major changes in government policy and economic theory.

> As the Industrial Revolution changed the labor conditions of Europe, it also changed the social, political, and economic structure of Europe forever.

In search of natural resources: Stealing is cheaper than dealing

The factories of the Industrial Revolution created wonderful products, but to do so they required natural resources. Europe had its share of coal and iron ore, used to provide power and equipment for the factories, but raw materials such as cotton and rubber had to be imported because they didn't grow in the climates of western Europe. As a result, the imperialist and industrial powers of Europe colonized regions of the world that could provide them with the raw materials they needed. This led to the policy of **mercantilism**. Mercantilist nations acquired incredible wealth by colonizing regions with natural resources, then taking those resources without compensating the natives and sending the resources back to Europe, where they were made into finished products. Then, the mercantilist nations sent those finished products back to the colonies, where the colonists had to purchase them because the colonial power wouldn't let the colonies trade with anyone else. In short, the colonial powers became rich at the expense of the colonies. The more colonies a nation had, the richer it became.

Soon, Europe was colonizing nations on every other continent of the globe. Europe became a clearinghouse for raw materials from around the world while the rest of the world increasingly became exposed to Europe and European ideas. What's more, the need for raw materials transformed the landscape of the conquered regions. Limited raw materials were being depleted faster than at any time in human history. Cotton plantations were spread across previously unfarmed land and after a few years, depleted the soil of its nutrients. Therefore, the unprecedented use of natural resources during the Industrial Revolution not only changed the lives of people on every continent, but the environment as well.

Economic thought: Capitalism gives way to socialism

Adam Smith wrote in *The Wealth of Nations* that economic prosperity and fairness is best achieved through private ownership. Individuals should own the means of production and sell their products and services on a free and open market, where the demand for their goods and services would determine their prices and availability. A **free-market system**, Smith argued, would best meet the needs and desires of individuals. Smith wrote *The Wealth of Nations* in 1776 in response to the western European mercantilist practices that had led to regulation of the economy. In the following century, as western Europe's economies developed under Smith's principles, industrialism gained momentum but also resulted in widespread exploitation of workers.

The Industrial Revolution created an enormous working class, unlike any class of people that came before. Factory employees worked long hours in cramped, dirty, and often dangerous environments, while the owners of the factories became wealthy. Unlike many classes of people who had been forced into labor conditions, the factory workers were told that this was their best option and that they, too, could rise to the top. **Karl Marx** did not believe that the factory workers had genuine opportunities. He believed that they were being exploited and that this exploitation was an inevitable consequence of capitalist industrialization. He wrote that the working class would eventually revolt and take control of the means of production, and while many western European workers attempted to do that, a communist revolution did not occur until 1917 in Russia. However, throughout the rest of Europe, social legislation was passed to improve the working conditions in the factories and establish minimum wages.

> Although very different in substance, the theories of Karl Marx and Adam Smith were both reactions to the Industrial Revolution.

White man's burden: New twist on the old ideas of exploitation

The **White man's burden** was an attitude of moral and social superiority of Europeans ("whites") over the natives of Asia, Africa, and the Americas. As you know, the Europeans invaded and colonized much of the world from the sixteenth century through the early twentieth century for a variety of reasons. Among the reasons were the desire to gain access to raw materials needed for industrialization; the desire to establish colonies that would purchase finished products from European industries; the desire to acquire land that would give a European power a strategic advantage over its adversaries; and the desire to be the largest, most powerful empire the world has ever known. *White man's burden*, however, characterized imperialism as a "burden." In other words, it justified imperialism as a responsibility of Europe. It is the duty of Europeans, Rudyard Kipling wrote in the poem "White Man's Burden", to conquer each "half-devil and half-child" so that those peoples can be converted to Christianity and civilized in the European fashion. If the natives refused, of course, they would be killed, but apparently that was in their best interest as well.

The poem, and the attitude that it nurtured, had a tremendous impact on public attitudes in Great Britain. As the British Empire expanded, the masses of Britain were also becoming more educated and sophisticated. They were aware that their government was not merely settling previously unsettled territories, but invading and taking land from people. The attitude of the poem, however, made these acts seem justified. Since they were told that the natives were "half-devil and half-child," the national feeling of superiority was enhanced.

The argument appealed to many people who were very closed-minded regarding the value of other societies, but yet genuinely wanted their nation to do what was morally right. As a consequence, many of them became missionaries to aid the imperialist effort. Since the poem gave imperialism a moral theme, Britain united behind imperialist conquest like never before. Britain was the undisputed world power of the early twentieth century. It possessed imperial holdings on every continent of the globe.

Changes in China

As you know, for much of its history, China was relatively isolationist, conducting only limited trade with other nations. As the West industrialized and European imperial powers expanded, trade with China became much more important. China was not only a vast land full of natural resources, but also a country with an enormous population that, though poor, attracted the attention of mercantilists who wanted to expand European markets. Up until the 1830s, China allowed the European powers to trade only in the port city of Canton, and it established strict limitations on what could be bought and sold.

However, Britain persistently tried to trade more than the Chinese government was willing to allow. Frustrated with the trading practices of the British, China seized opium that British merchants were trying to sell, starting the **Opium Wars** that lasted for the next decade. Britain, with its superior military might, overwhelmed China and consequently acquired access to many other ports, opened up Chinese markets, and made territorial acquisitions, including the land that would become Hong Kong. When other European powers saw how easily Britain gained access to China, they rushed with ships to get a piece of the pie. The European powers subsequently carved up China into spheres of influence, leading to widespread discontent and humiliation for the Manchu Dynasty. Then other Asian powers realized how weak China had become. The Koreans revolted against Chinese rule, and the Japanese actually attacked. All of this didn't sit too well with the Chinese peasants.

The Society of Righteous and Harmonious Fists, or **Boxers**, organized in response to the Manchu government's defeats and concessions to the Western powers and Japan. The concessions of China were highlighted by two major events: first, the **Treaty of Nanjing**, which involved China's release of Hong Kong to Great Britain; and second, the **Treaty of Shimonoseki**, which involved Japan's acquisition of Taiwan. Infuriated, the Boxers rose in the 1890s to drive the Europeans and Japanese out of China. Adopting guerrilla warfare tactics, the Boxers slaughtered Christian missionaries and seized control of foreign embassies. Ultimately, however, they were not successful in achieving their goals. Instead, their uprising resulted in the dispatching of foreign reinforcements who quickly and decidedly put down the rebellion. The Manchu government, already having made great concessions to the Europeans and Japanese, was now even further humiliated. As a result of the rebellion, China was forced to sign the **Boxer Protocol**, which demanded indemnities to the Europeans and Japanese for costs associated with the rebellion. Thus, the Boxers were not only unsuccessful in reclaiming China under nationalistic control, but also ultimately responsible for increasing the foreign presence in China.

> Although it ultimately failed, the Boxer Rebellion would provide inspiration for the first Republic of China.

Eventually, the vast majority of the citizens became angry with both the foreigners and the Manchu Dynasty, and revolted under the leadership of Sun Yat-sen, and established the first Republic of China in 1911. There is more information on this topic later in the chapter.

Changes in Africa

Prior to the discovery of gold and diamonds in South Africa in the 1860s and 1880s, South Africa was valuable to the Europeans only for shipping and military reasons. The Dutch arrived first and settled Cape Town as a stopping point for ships on the way from Europe to India. In 1795, the British seized Cape Town for themselves, and the South African Dutch (now known as *Boers* or *Afrikaners*) trekked northeast into the interior of South Africa, settling in a region known as the Transvaal. When the Boers later discovered diamonds and gold in the Transvaal, the British followed the Boers into the interior and fought a series of wars against them for rights to the resources. After years of bloody battles, the British reigned supreme, and all of South Africa was annexed as part of the ever-expanding British Empire. Thus the discovery of valuable gold and diamond mines greatly affected the European conquest of the region. Of course, throughout this entire process, the natives were not allowed claims to the gold and diamonds, and were made to work in the mines as their natural resources were sent abroad.

Meanwhile, in the rest of Africa during the nineteenth and twentieth centuries, European imperialism was not accepted willingly by most native Africans. As a result, conflicts between African nationalists and colonial rulers became increasingly common as time went on. There was much to be resented in the way the Europeans imposed themselves on Africa. The European nations of Belgium, Britain, France, Germany, Spain, Portugal, and Italy had carved up the entire continent of Africa into colonies, without regard for the original tribal loyalties. Instead, they agreed on boundary lines at a conference in Berlin in 1845 and drew the lines solely based on bargaining for political and economic advantage. The results led to chaos: In some situations, tribal lands were cut in half between two colonies controlled by two different European nations, while in other situations two rival tribes were unwillingly brought together under the same colonial rule. For a time, the disruption of the traditional tribal boundary lines worked to the Europeans advantage since it was difficult for the native Africans to organize an opposition within each colony.

Changes in Japan

During the Industrial Revolution, the United States exerted pressure on Japan to open its ports of trade by dispatching people to Japan. For example, **Commodore Matthew Perry** arrived on a steamboat (something the Japanese had never seen before) and essentially freaked out the Japanese. The Japanese realized that their isolation had resulted in their inability to compete economically and militarily with the industrialized world. For a time, the West won concessions from Japan through various treaties such as the **Treaty of Kanagawa**. These treaties grossly favored the United States and other countries. As in China, the nationalists grew resentful, but unlike the Chinese, the Japanese were organized. Through the leadership of the samurai, they revolted against the shogun, who had ratified these treaties, and restored Emporer Meiji to power.

The Meiji Restoration ushered in an era of Japanese westernization, after which Japan emerged as an aggressor. Rather than falling victim to colonization plans, Japan joined in the game and colonized parts of Asia for access to raw materials needed for its own developing industrial base.

> The term *Weternization* is almost synonymous with *modernization*—it means changing attitudes so that they more closely reflect the Western world's emphasis on individualism and technology.

Changes in Russia

During the Industrial Revolution, Russia was thrust into war with Japan when Japan attacked the Russian fleet at Port Arthur in Manchuria. Both Russia and Japan had been vying for control of Manchuria, which was rich with natural resources. Russia's loss to Japan was humiliating. For the first time, a modern European power had lost to an Asian power, and as an added insult, had been defeated with Western weaponry and technology. This not only sent a wake-up call to Moscow, but to the entire world. Japan was emerging as a world power. As for Russia, it lost land and fishing rights to Japan, but internally, the government lost much more. The czar lost legitimacy, setting the stage for insurrections and eventually revolution.

Quick Quiz 7

1 **In the nineteenth century, the unification of Italy and the unification of Germany resulted in**

A upsetting the balance of power in Europe

B increasing competition for trade with Russia

C reducing nationalism in these nations

D encouraging a century of peaceful coexistence in Europe

2 **A major cause of the Opium War in China was**

F an increase in the power of the emperor

G the establishment of spheres of influence in China by Europeans

H the expansion of Chinese influence to India and the Middle East

J the expulsion of Europeans from China

3 **In an attempt to modernize Japan during the late 1800s, the leaders of the Meiji government decided to**

A study Western institutions and technology

B maintain a policy of isolationism

C establish close relations with China

D end the political power of the Buddhists

4 **According to the theory of mercantilism, colonies should be**

F acquired as markets and sources of raw materials

G considered economic burdens by the colonial power

H granted independence as soon as possible

J encouraged to develop their own industries

5 **The nineteenth century term white man's burden reflects the idea that**

A Asians and Africans were equal to Europeans

B Asians and Africans would be grateful for European help

C imperialism was opposed by most Europeans

D Europeans had the responsibility to colonize the rest of the world

Check your answers to Quick Quiz 7 on page 153. If you missed any of these questions, reread the previous above.

Twentieth-Century World Conflicts

Most of the questions about the twentieth century will deal with major world conflicts and the end of European imperialism. This section is organized as follows:

- World War I: The war to end all wars?

- The Russian Revolution

- Totalitarian men: Mega-jerks of history

- World War II: Here we go again

- Cold War: Let's point our weapons at each other

- Revolution in Asia: Communist gains

- The end of colonialism: Free to be me

- Remaining religious conflicts

There's a lot here to digest, and there will be 10 related questions on the exam, so review carefully.

World War I: The War to End All Wars?

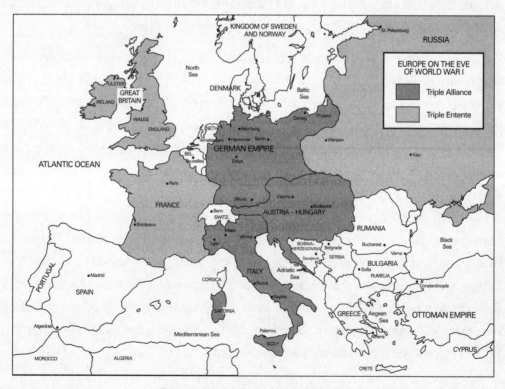

Europe on the Eve of World War I

When **Archduke Ferdinand of Austria-Hungary** was assassinated by a Serbian nationalist in Sarajevo, Austria-Hungary made demands on the Serbian government. Although the Serbian government complied with most of the demands, Austria-Hungary was not satisfied. It wanted absolute control over the Balkans, which included Serbia. So it joined forces with Germany and they attacked Serbia, setting off World War I. Russia sided with Serbia; Germany attacked France; France sided with Russia; Turkey sided with Austria-Hungary; Great Britain sided with France and millions of Russians and Europeans lost their lives.

The consequences for Austria-Hungary were devastating, so devastating that Austria-Hungary ceased to exist. Instead, it split into three nations: Austria, Hungary, and Czechoslovakia. As in many other European nations, millions of lives were lost, especially men in their teens and twenties. In short, Austria-Hungary's aggressive response to the assassination cost it its existence.

As for the rest of Europe, it was devastated. The war was so huge that, at the time, it was called The Great War or The War to End All Wars. Weaponry used included grenades, poison gas, flame throwers, and machine guns (the Industrial Revolution served artillery engineers well, unfortunately), and the casualties were enormous. Russia was so shattered that it pulled out even before the war was over—they had their own internal revolution to focus on (more on that later). After Russia pulled out, Germany thought it would have an easier time defending itself, but then the United States entered the war in 1917, and it was just too much for the Germans. The fighting finally ended on November 11, 1918.

But the important thing about World War I isn't the war itself, but the way it was resolved. The **Treaty of Versailles**, signed in 1919, brought an official end to World War I. It was extremely punitive against Germany, which was required to pay war reparations, release territory, and downsize its military. The treaty represented a departure from U. S. President Wilson's **Fourteen Points**, which Germany had originally agreed to when it signed the armistice, but which the allies of the United States did not consider sufficient. Britain and France, for example, needed to justify the human and financial costs of the war to their own demoralized citizenship. So the victors blamed the war on Germany and then, over the objections of the United States, forced Germany to sign an extremely punitive treaty. The victors hoped that as a result of the treaty, Germany would never threaten the security of Europe again. Instead, the treaty caused widespread economic problems and resentment in Germany. As a result, a fascist nationalist, Adolf Hitler, was able to rise in power, promising restoration of the German Empire. Therefore, the Treaty of Versailles ended one war, but it led to a much more disastrous war just two decades later.

Another major consequence of World War I was the rise of Ataturk and the establishment of the modern Republic of Turkey (an area previously at the center of both the Byzantine and Ottoman Empires). **Ataturk**, "the Father of the Turks," was the first president of modern Turkey. He earned the adoration of millions of Turks during World War I and then used that loyalty to gain support in his political career. He successfully secularized the overwhelmingly Muslim nation, introduced Western dress and customs (abolishing the fez, a traditional type of hat), changed the alphabet from Arabic to Latin, set up a parliamentary system (which he dominated), changed the legal code from Islamic to Western, and set Turkey on a path toward westernization.

However, Ataturk instituted these reforms sternly against opposition and sometimes was ruthless in his determination to institute change. Nevertheless, he succeeded where so many others would have failed. The economy and opportunities in Turkey are still growing today as a result of his reforms, despite growing Muslim fundamentalism.

> World War I is considered the first modern war because it involved many nations and brand-new, incredibly destructive military technology.

The Russian Revolution

Weakened by World War I, scandals, and popular uprisings for basic necessities, the **Romanov Dynasty** was coming to an end by the second decade of the twentieth century. When his troops would no longer follow his orders, **Czar Nicholas II** abdicated the throne, sending the country into revolution. In March of 1917, moderates took control by setting up a Provisional Government to fill the void left by the Romanovs. They intended to write a constitution and establish a democracy, but they weren't willing to institute major economic changes that would have brought some real relief to the suffering peasants. What's more, the moderates kept Russia in World War I, which angered most of the peasants.

So in November, radicals, led by **Vladimir Lenin** and **Leon Trotsky**, formed the Bolshevik Party and with the help of some of the peasants, overthrew the Provisional Government and killed the Romanovs. The **Bolshevik Revolution** (sometimes called the October Revolution) intended to create a Marxist, or Communist, state, in which everyone would be equals, no one would be exploited, and eventually, if everyone contributed to society based on their abilities, there would be no need for government. Or so the theory went. Before this could be accomplished, the Bolsheviks reasoned, they would have to control Soviet society until evil could be rooted out.

Vladimir Lenin, the leader of the Bolshevik Revolution and the first dictator of the Soviet Union, believed that a strong central government with absolute power was necessary until society eliminated capitalism, organized religion, and any other force that he believed threatened the people. In other words, he believed that the officials of the Communist Party knew what was best for the people, and therefore, opposition was not tolerated. Lenin did not believe that the authority of his dictatorship was limited by natural rights or even by Communist philosophy. He believed that strong leadership was the primary force that secured those rights. Therefore, even though Lenin's goal was to achieve a perfect Communist society, in an effort to achieve that goal he engaged in a variety of methods that were not consistent with pure communist philosophy, including exercising strong central authority and, at times, allowing for limited capitalism to spark the economy. Under his leadership, the Soviet Union became one of the largest and most powerful countries on Earth.

Totalitarian Men: Mega-jerks of History

Totalitarian dictators, also known as fascists, rule as a single, absolute authority. They often come to power by way of a military coup or revolution. Totalitarian dictators generally do not simply make laws but rather attempt to control society in very fundamental ways. They often ban opposition entirely, grossly violate basic human rights, and try to control mobility and even the dissemination of information within their boundaries. In addition, totalitarian dictators are often aggressive against their neighbors. They look to obtain not only more power, but also more power over more people.

Stalin: Mega-jerk #1

The Soviet Union under **Joseph Stalin** is an example of a totalitarian dictatorship. Stalin was able to use his authority to mobilize the nation in his **Five-Year Plans** for economic development, but all of the negative aspects of a totalitarian dictatorship were also present. Stalin became dictator of the Soviet Union after Lenin's death in 1924. He imposed his will ruthlessly. Those who opposed him often were killed. After collectivizing agriculture, Stalin built up the nation's heavy industry under his Five-Year Plans. The Soviet people went without many of the products that the rest of the industrial world was enjoying while the Soviet Union built up its military and industry. Nevertheless, the development came just in time to allow the Soviet Union to prevent occupation by Germany in World War II. Under Stalin's leadership, the Soviet Union emerged from World War II as one of the world's two superpowers. But in the process, millions were killed by his secret police, and individuality was stifled completely. To this day, Stalin is widely regarded as one of history's most ruthless dictators.

Hitler: Mega-jerk #2

The other big mega-jerk totalitarian from this time period was **Adolf Hitler**. Hitler was a man of very dramatic speech who often made strong appeals to German nationalism, which he brought to a boiling point by not only reminding Germans how they had been wronged by other nations but also advancing the idea that Germans were superior to all other peoples.

These ideas were first stated by Hitler in ***Mein Kampf***, his personal account of his struggle, which later became the definitive statement of the goals of the **Nazi Party**. In the 1930s, the Nazis became the largest political party in Germany by asserting the claims in Hitler's book— most notably that Germany had been wrongly victimized by the Treaty of Versailles after World War I, that the Jewish people and other "undesirables" were behind many of Germany's problems, and that the Aryan Race was destined to rule Germany and the world.

The goal of the party, therefore, was to gain control of Germany and establish an Aryan superpower. Under the leadership of Hitler, who was sworn in as chancellor in 1933 and then assumed the title and authority of *Fuhrer* of the Third Reich, the Nazis swiftly created a new political and social order. They utilized a secret police force, the **Gestapo**, to quash dissent. They built concentration camps to carry out genocide against the Jewish people, Gypsies, homosexuals, and others (an atrocity known as the **Holocaust**). And all the while, they united Aryan Germany by feeding the flames of Aryan nationalism. Ultimately, however, the Nazi defeat in World War II served as a lesson to the world on the dangers of ethnocentric nationalism that seeks to elevate itself at the expense of others.

> Although Hilter was a totalitarian dictator, he was actually elected as Chancellor of Germany. Shortly thereafter, he removed all traces of democracy from the country.

World War II: Here We Go Again

Egomaniacs playing with bombs

The events that led to World War II go back to World War I. In fact, in a very real way, WWI and WWII are pretty much the same war, just part one and part two. Recall that under the 1919 Treaty of Versailles, Germany was severely punished by the victors of World War I, who forced Germany to give up much of the territory it prized. The newly created nation of Czechoslovakia was one of the beneficiaries of the treaty. Land that had been occupied and settled by Germany was handed over to Czechoslovakia, essentially "trapping" millions of Germans within the boundaries of a new nation. As Hitler's Nazi party aroused nationalism and Aryan racism, Germany wanted to reunite all previous German territories. The European powers made concessions to Hitler and allowed him to take control of the Czech territories of Bohemia and Moravia, but Hitler wasn't satisfied. Instead, Germany attacked Czechoslovakia in 1939 and then moved on to take control of Poland. The Allies finally stopped making concessions and World War II had begun. Europe was at war with itself once more. Germany allied itself with Italy, which also was led by an ethnocentric totalitarian jerk named Mussolini, while Great Britain, France, and the Soviet Union were allied together to stop them.

Already in a war with China, Japan entered World War II on December 7, 1941, when it attacked the United States at Pearl Harbor. World War II had already consumed Europe by this time, and Japan had allied itself with Germany and Italy, but it was Japan's interest in China that brought it into the war. Japan wanted to overrun China and gain access to its resources. The United States made it clear that it would not allow that to occur and imposed an economic embargo on Japan. In response, Japan attacked Pearl Harbor and declared war on the United States. When Germany, Japan's ally, also declared war on the United States, Japan became part of a much larger war than just its war with the United States and China.

The consequences of the war transformed Japan. Japan was militarily destroyed, and in the end, two of its cities, Nagasaki and Hiroshima, were destroyed by atomic bombs dropped by the United States. Its imperial empire was stripped away, its military was reduced to almost nothing, and its government system was transformed into a parliamentary democracy. After a crushing defeat, Japan recreated itself in the political and economic mold of the West.

In Europe, the consequences were enormous as well. Millions dead. Untold damage. It was a mess. The victorious allies were hardly in the mood for celebrating because they had to rebuild entire countries. What's more, Great Britain, France, and the Soviet Union didn't particularly like each other, so establishing peace on the continent was going to be quite difficult.

Aftermath

Here's a quick summary of what happened after World War II: Poland got more land. Italy returned to democracy after overthrowing Mussolini. Communism spread to Eastern Europe under the influence of the Soviet Union, while democracy was entrenched in Western Europe under the influence of the United States. Nuclear proliferation expanded as everybody wanted the powerful bombs the United States had used against the Japanese.

The **United Nations** was established to promote world peace, though it has no formal power to force a country to comply with its determinations. More significantly, two new organizations entered the world stage that would set the tone for the balance of power for decades to come: the **Warsaw Pact** (dominated by the Soviet Union) and the **North Atlantic Treaty Organization** (**NATO**, dominated by the United States). These two organizations were pitted against each other in what has become known as the **Cold War**.

Quick Quiz #8

1 **Which idea was included in the Treaty of Versailles to show the intent of the Allies to punish the Central Powers for their role in World War I?**

  A All nations shall maintain open covenants and peace.

  B Freedom of the seas will be maintained.

  C Germany will accept full responsibility for causing the war.

  D Territorial settlements shall be made along clearly recognizable lines of nationality.

2 **Fascism in Europe during the 1920s and 1930s is best described as a**

  F demonstration of capitalism

  G form of totalitarianism

  H classless society

  J set of humanist ideas

3 **Under Joseph Stalin, life in the Soviet Union was characterized by**

  A an abundance of consumer goods

  B political instability and numerous civil wars

  C support for small family-run farms

  D the use of censorship and the secret police

4 **The Cold War that emerged at the end of World War II was between**

  F the United States and China

  G the United States and the Soviet Union

  H the Soviet Union and China

  J Germany and the Soviet Union

Check your answers to Quick Quiz #8 on page 153. If you missed any of these questions, reread the previous section.

The Cold War in Europe

Cold War: Let's Point Our Weapons At Each Other

NATO and the Warsaw Pact: A staring contest that lasted decades

After World War II, the United States and the Soviet Union emerged as the world's two big superpowers. One was capitalist and a democratic, the other communist and totalitarian. Each wanted to remake the world in its own image. Both couldn't stand the other but were afraid of the consequences of all-out war. So, instead of fighting, they built up their militaries, courted every other country on the planet to ally with, and stared each other down, trying to stop the other from expanding their influence.

As a consequence, Europe was literally split in half. **NATO**, led by the United States, established itself as a military alliance of Western European powers, as well as Greece and Turkey. The **Warsaw Pact**, led by the Soviet Union, established itself as an alliance of Eastern European powers, such as Poland, Czechoslovakia, and Romania. As for Germany, the war-torn country was torn in half: West Germany for NATO, East Germany for the Warsaw Pact. Even the city of Berlin itself, located in East Germany, was split into two. The division between the two alliances was so severe that Winston Churchill declared there was an **Iron Curtain** splitting Europe into two. For nearly 40 years, the two superpowers built up their military might along the Iron Curtain, both sides claiming they were doing it for defensive purposes.

The Cold War turns hot in Asia

While the two superpowers didn't pull the trigger in Europe and provoke the other, they seemed to have no problem playing out their feud in Asia, where the Cold War sometimes turned hot. Korea was the stage for the first big conflict.

Prior to World War II, Korea was invaded by Japan and annexed as part of the expanding Japanese Empire. After Japan was defeated in World War II, Korea was supposed to be reestablished as an independent nation, but until stability could be achieved and elections held, it was occupied by the Soviet Union and the United States in two separate pieces—the Soviet Union north of the thirty-eighth parallel and the United States south of it. As the Cold War developed between the two superpowers, they couldn't agree on the terms of a united Korea. In the meantime, each superpower continued to exert its own influence over the political and cultural developments within the territory it occupied. In 1948, two separate governments were established: a Soviet-backed communist regime in North Korea and a U.S.-backed democracy in South Korea. Both superpowers withdrew their troops in 1949, but in 1950, North Korea attacked the South in an attempt to unite the two nations under a single communist government.

> After Korea, the Cold War spilled into Vietnam.

The United Nations condemned the action and soon a multinational force, largely consisting of U.S. and British troops, went to the aid of the South Koreans. These United Nations forces made tremendous headway under **General Douglas MacArthur**, but when it looked as if the North Koreans would be defeated, China entered the war on behalf of the communist North. The two sides battled it out along the thirty-eighth parallel, eventually leading to an armistice in 1953. Today, the two nations remain separate and true to the political philosophies under which they were created, although very recently there has been diplomatic communications between the two sides.

Ho Chi Minh was the founder of the Socialist Republic of Vietnam. He and his communist followers drove Japan from Vietnam, and then prevented the French from reoccupying the nation after World War II. An accord signed in Geneva in 1954 divided the nation into two. The communists, under the leadership of Ho Chi Minh, gained control of the land north of the seventeenth parallel while **Ngo Dihn Diem** became the president of the democratic South. Under its new constitution, North Vietnam supported reunification of Vietnam as a communist state. Ho Chi Minh supported communist guerrillas, known as the **Viet Cong**, in the South, and soon war broke out. France and the United States came to the aid of South Vietnam, but Ho Chi Minh's Viet Cong prevented them from securing a victory. A peace agreement eventually led to the reunification of Vietnam as a communist state.

The Cold War finally thaws

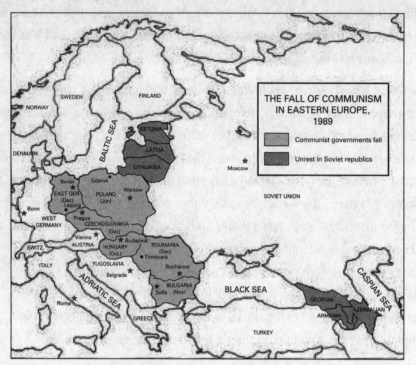

The Fall of Communism in Eastern Europe

In the early 1990s, dramatic changes took place. Poland started to break away from the Soviet Bloc; the West was clearly on the technological fast track; and reforms were instituted in the Soviet Union that soon could not be reversed.

Mikhail Gorbachev, the last leader of the Soviet Union, instituted widespread economic and political change, leading to the downfall of communism in the region. Under his plans of *perestroika* and *glasnost*, Gorbachev restructured the Soviet economy to include elements of free-market capitalism and opened Soviet society to the world community, allowing for the free flow of information and political ideas. The reforms were intended to spark the stalling economy and nurture innovation, but they soon took on a life of their own. Increasing numbers of people wanted not only reform, but also revolution. Soon, the Soviet Bloc dismantled as nations wanted to chart their own course under democracy and capitalism. In addition, the Soviet Union itself dismantled into independent republics.

Since the Soviet Union no longer existed, Gorbachev no longer had a nation to lead, but his successor, **Boris Yeltsin**, continued the reforms as leader of the Republic of Russia. Many hard-line communists oppose the changes that have occurred, and the future remains uncertain as the former communist nations struggle to adapt to free-market competition. However, the impact on the region, the world community, and the future of communism is undisputed. Gorbachev set forces in motion that have changed the economic and political landscape of the world.

> Today, the United States and Russia have a cooperative relationship, and the Cold War has essentially melted.

Revolution in Asia: Communist gains

The **Manchu Empire** was the last dynasty of China. In the early twentieth century, **Sun Yat-sen** challenged the dynasty that had ruled in China since the seventeenth century. He claimed that their rule had led to unwanted foreign intervention and a lack of social and economic development. Previous to Sun Yat-sen's Chinese Revolution of 1911, the peasants generally worked on small subsistence farms and had little need for education or exposure to the outside world. You should recall that they were exposed nonetheless due to invasions from Japan and the establishment of European spheres of influence. Sun Yat-sen wanted to change all of this with his **Three Principles of the People**: nationalism, socialism, and democracy. Nationalism would unite the people against foreign interests and give them a Chinese identity; socialism would lead to greater economic equality and opportunity; and democracy would lead to the ability of the Chinese people to chart their own future.

Dynastic rule ended when Sun Yat-sen established the **Republic of China** in 1911; however, most of his intended reforms were not carried out. Therefore, Sun Yat-sen's changes were important in that they ended dynastic rule, but they failed to replace the void with a workable system. Later, the government of **Chiang Kai-shek** made attempts to continue the policies of Sun Yat-sen by working through a representative political party in South China known as the **Koumintang**. Chiang Kai-shek befriended the Western powers and served as a barrier to the growth of communism in the region, especially after he unified the northern part of China with the southern part, keeping the nation on track to become a liberal democracy.

However, during this time, imperialist Japan attacked Manchuria in northern China, and while the government of Chiang Kai-shek dealt with the invasion, the communist party under **Mao Zedong** gained support. Mao believed that the government was ignoring the basic needs of millions of peasants. After building up a rural peasant force of over 1 million soldiers, Mao Zedong swept through China and drove the Koumintang off mainland China and over to the nearby island of Taiwan, where the Koumintang created the Republic of China.

Throughout his life, Mao strongly adhered to basic communist principles and tried to expand the influence of communism throughout Asia (see the discussion of Korea and Vietnam above). But it was in China itself that Mao totally transformed life. As part of his **Great Leap Forward** and **Cultural Revolution**, Mao sent millions of peasants, government bureaucrats, and university teachers to work side-by-side in collectivized farms; kept strict control of all means of production; remained intolerant of dissent; and ruled as a dictator over virtually every aspect of Chinese life. Even after Russia started to back away from its aggressive communist agenda, China remained the world's largest and most committed communist nation. Even today, China remains one of the few communist countries, although in recent decades, it has embraced elements of capitalism and is poising itself to enter the world's economic stage in a huge way.

> China remains a communist country, but it is also one of the biggest emerging economic powers in the world.

The End of Colonialism: Free To Be Me

Israel emerges after the war

The state of Israel was created as a Jewish homeland in 1948 under the authority of the United Nations in a small region along the eastern edge of the Mediterranean known as Palestine. For decades prior the creation of Israel, thousands of Jewish people had already been moving to the region as part of the Zionist movement, an organized effort by the Jewish people to settle the land that was promised to them by God through Moses. When the nation of Israel was formally created, the Jewish people rushed from Europe and the United States to establish the new nation.

However, this land was previously occupied by Muslims, who also claimed it as their own. The Muslim nations that surround Israel refused to recognize it as a Jewish homeland. Soon violence erupted and eventually wars broke out, but Israel held its ground and expanded its territory. The **Palestinian Liberation Organization (PLO)** formed to liberate the region from the Jewish people and sponsored violence against Israel. Although recent agreements between Muslims and the Jewish people have offered hope for peace in the region, violence persists. The creation of the state of Israel and the Muslim reaction to it, therefore, have created a tremendously unstable political, religious, and military environment in the entire Middle East region.

India

Britain controlled the Indian subcontinent in the nineteenth and early twentieth centuries, but resentment grew stronger among the Indians as colonialism wore on. The British started to worry that there might be a popular uprising in India. After the success of the Bolshevik Revolution in Russia, Europe was alarmed. Britain became suspicious of any nationalistic tendencies within its colonies, most especially its crown jewel of India. Britain instituted tough new laws against Indian conspiracy for nationalism, and this, of course, only provoked more nationalism. It was at this point that **Mohandas Gandhi** first encouraged a change in British policy through peaceful civil disobedience, but some of his countrymen were more violent. They attacked and killed a few Englishmen in Amritsar in 1919. A British general reacted against the crowd, not just against the individual aggressors, and killed nearly 400 Indians, injuring hundreds more.

The consequences of the Amritsar revolt and massacre were not immediate, but they were significant. Many British officials defended the general, while Indians rallied behind the nationalistic cause. In the aftermath of this violence, Gandhi maintained his practice of peaceful resistance, which allowed him to gain worldwide legitimacy and support. The British were reluctant to pursue violent suppression of nationalism against a well-maintained nonviolent opposition.

When Gandhi spoke of the consequences of British imperialism, he spoke not only from a nationalist perspective with regard to his native India but also from a global perspective. Born in India, Gandhi was educated in Britain and spent several years in British South Africa, where he saw first hand the consequences of institutionalized racism. Outraged, he became an advocate for Indian nationals in South Africa and then returned to India, where in the 1920s and 1930s, he embarked on a large-scale resistance against British rule, leading such efforts as the **Salt March** and the **Quit India Movement**. Although he spoke with great determination and refused to back away from his goals, he was strongly opposed to the use of violence. Instead, his approach is one of history's clearest examples of civil disobedience. He opposed British imperialism, attempted to transform rural education, attempted to improve the rights of members of India's lowest caste, and encouraged Hindu-Muslim unity, all through peaceful measures aimed at solving problems rather than contributing to them.

When Britain finally granted India its independence, there were two plans proposed. The first was a movement by Gandhi to establish a united India where both Hindus and Muslims could practice their religions. The second was a movement by **Muhammad Ali Jinnah** to form a separate Muslim nation in the northern part of the Indian subcontinent, where Islam had become the dominant religion. After World War II, Britain finally granted independence to the Indian subcontinent but separated it into thirds: India in the south, and Pakistan in two parts, one northwest of India and the other east. Both parts of Pakistan were Muslim, while India was predominately Hindu, although officially secular. The result was chaotic. Millions of people moved or were forced to flee due to religiously motivated violence. Es-

> Gandhi's principles of peaceful protest and passive resistence have been used by millions of groups throughout the world to furthur their causes.

sentially, India and Pakistan exchanged millions of citizens, with practitioners of each religion moving to the nation where their religion was dominant. Eventually, East Pakistan became Bangladesh but remained primarily a Muslim nation. However, the move of so many people along religious lines served only to fuel the conflict between Pakistan and India. Today, the two nations are still fighting, especially in the region of Kashmir along their borders, where religious self-determination still remains the big issue.

Sub-Saharan Africa

By the beginning of the twentieth century, nationalism among native Africans grew intense. The Africans resented the way in which the Europeans imposed their own culture, language, and religion. They resented that they were treated as second-class citizens at best, and slaves at worst. They resented that they were not represented in the colonial governments. Yet, since most of the resistance was relatively peaceful, it became more and more difficult for European governments to ignore the calls for independence, and international pressure to grant Africans the right to self-determination increased, especially in the years during and immediately following World War II. By 1945, major independence movements were underway.

In Kenya, for example, **Jomo Kenyatta** encouraged feelings of nationalism that led to protests against Europeans as well as Asians, who also had settled in Kenya. At the same time, guerrilla groups known as the **Mau Mau** began to instigate violence against the white settlers. Kenyatta was accused by the British of leading the group and was sentenced to seven years of hard labor, but the violent uprisings continued. The British finally abandoned their attempts to suppress the uprisings and granted independence in 1963, at which point Jomo Kenyatta became president of the new nation of Kenya. Kenyatta did not look to reunite his nation after the bloody violence. Instead, he ordered the removal of whites and Asians, which had an immediate adverse impact on the economy. Kenyatta persisted his reforms, however, and Kenya's economy grew swiftly.

Overall, peace and prosperity in the newly independent countries of Africa has been a challenge to maintain. Part of the problem stems from the way in which the European powers carved out their colonies and then carved out the newly independent countries. In both cases, the Europeans drew the boundaries of new nations according to their own political motivations. No consideration was given to the identities of the people within the boundaries. Tribes that were formerly united often were split into two, while rival tribes often were grouped together within the same nation.

As a result, the governments of many African nations have had difficulty establishing a truly "national" purpose and identity. **Tribalism**, rather than nationalism, has been the driving force behind many community goals. Many African political leaders, such as **Kwame Nkrumah**, recognized this problem immediately. As the first president of Ghana, he attempted to unite the nation under common goals, but divided public opinion and mounting national debt made him assume even more centralized authority in an attempt to enforce his plans. He was eventually overthrown in a 1966 military coup.

To complicate matters further, the effect of imperialism on the continent of Africa remains enormous. Most African nations have structured their governments on the European models, have adopted their colonial languages as official languages, and have in most cases retained the colonial boundary lines. In addition, the prevalence of Christianity in sub-Saharan Africa also stems from the colonial period.

> Despite setbacks in in the past, nationalism in Africa continues to grow as Africans rediscover their pasts while also preparing for the future as powerful, independent, and sovereign nations.

South Africa

South Africa has a history that is distinct from that of the rest of the continent. Dutch settlers originally arrived in the 1600s, but by 1815, Holland agreed to give up control of its South African colony to the British, who were interested in it for naval reasons. Unwilling to be governed by the British, however, the Dutch settlers, or **Boers**, moved north into the interior of South Africa, away from the British settlements along the coast. When gold and diamonds were discovered in the interior, the British went to war with the Boers for control of the region. By 1910, the British had control of the entire region and established it as the Union of South Africa.

In 1948, the Boers (also known as Afrikaners) outnumbered the British and gained control of the government. They established a policy of **apartheid**, which literally means "apartness." Blacks comprised nearly 70 percent of the population and yet were denied citizenship and access to public facilities. Separate regions were set aside for them, but only the worst land was made available.

The **African National Congress** (ANC), which was formed to advocate rights for Black Africans, elected **Nelson Mandela** as its leader. The anti-apartheid movement gained momentum through the 1950s. The **Sharpeville demonstrations** was a part of the effort to gain support for the anti-apartheid cause. It was a disaster. The police opened fire on the demonstrators in Johannesburg, killing 72. A state of emergency was imposed, over 20,000 blacks were arrested, and the African National Congress was banned.

The massacre only fueled the anti-apartheid cause, however. Mandela continued his resistance and was arrested just a few years later, remaining in prison until 1990. The massacre also enraged much of the Western world. While Western powers were reluctant to intervene in South Africa on behalf of the black natives, world opinion was turning against the South African government after the massacre. Eventually, the goal of the Sharpeville demonstrations was realized when the South African government abolished apartheid in 1991. In 1994, South Africa's first multiracial election resulted in Mandela's presidency.

Remaining Religious Conflicts

Christian conflict in Ireland

The **Irish Republican Army (IRA)** is a group of Catholics in Northern Ireland, a part of Great Britain that is predominately Protestant. When Britain granted Ireland independence, it maintained control of six northern provinces known as Northern Ireland. This seemed to be a good idea at the time because Ireland was predominately Catholic while Northern Ireland was predominately Protestant, like Britain. However, the government of Northern Ireland discriminated against the Catholic minority, and eventually the IRA formed to combat the discrimination.

The IRA wants to join Ireland, while the Protestant majority wants to remain a part of the United Kingdom. Civil rights protests by Catholics in the late 1960's led to violence, including bombings, and soon both sides were inflicting violence against each other. The consequence of the IRA insurrections is not yet known. We do know that the conflict has raged on for decades and that thousands have died. We also know that both sides have been willing to negotiate in recent years. But since the dispute remains unsettled, and since the Catholic population of Northern Ireland is quickly increasing in proportion to the Protestant population, the future of Ireland and the success or failure or the IRA insurrections are still in question.

Fundamentalism in Iran

During the 1960s, **Shah Pahlevi**, a leader intent on westernizing Iran, instituted land and education reform, and increased the rights of women. As time passed, the shah infuriated many Islamic fundamentalists who wanted to make the teachings of the Koran the law of the land. They did not like the economic and social changes and believed that the influence of the West was too strong. Others believed that the shah was not reforming *enough*, especially concerning the political system, which lacked significant democratic reform. The shah reacted violently against dissent from both sides, pressing forward with his own mix of social and economic reform even in the face of strong public opposition. When President Jimmy Carter of the United States visited Iran to congratulate it on its programs of modernization and westernization, the Islamic fundamentalists had enough.

In 1979, the shah was ousted from power during the **Islamic Revolution**, which reverted Iran back to a theocracy led by **Ayatollah** ("mirror of God") **Khomeini**. Immediately, modernization and westernization programs were reversed, women were required to wear traditional Islamic clothing and to return to their traditional roles, and the Koran became the basis of the legal system. Since the death of Khomeini in 1989, the opposition to Islamic fundamentalism has gained momentum, especially in the area of economic reform. Today, the economic, social, political, and religious future of Iran is hotly debated.

Quick Quiz #9

1 World War II is often considered a turning point in history because

    A the League of Nations demonstrated that an international organization could maintain world peace

    B the war brought an end to dictatorships as a form of government

    C European domination of the world began to weaken as nationalism in the colonies increased

    D religious and ethnic differences were no longer a source of conflict between nations

2 During the Cold War, the United States and the Soviet Union were reluctant to become involved in direct military conflict, especially in Europe, mainly because of

    F the peacekeeping role of the United Nations

    G pressure from nonaligned nations

    H the potential for global nuclear destruction

    J increased tensions in the Middle East

3 Mohandas Gandhi is best known for his

A use of passive resistance to achieve Indian independence

B desire to establish an Islamic nation

C opposition to Hindus holding political office

D encouragement of violence to end British rule

4 One characteristic of apartheid, which was practiced in South Africa, is

F forced migration of blacks to other nations

G integration of all races in society

H an open immigration system

J segregation of the races

5 Which of the following is associated with Mao Zedong?

A the Cultural Revolution

B the Bolshevik Revolution

C the French Revolution

D the Industrial Revolution

Check your answers to Quick Quiz #9 on page 153. If you missed any of these questions, reread the previous section.

History Skills

History skills questions will require you to apply the content of world history (as reviewed above) and analyze and apply it in a variety of ways. There will be seven history skills questions on the exam. You will be expected to understand and use:

- Primary and secondary sources

- Trends

- Time lines

- Distribution

Each one of these skills is discussed below.

Primary and Secondary Sources

The test writers will include several questions on the exam that will involve a quotation from a primary source or a secondary source. A primary source is a person who is directly involved in the historical situation. For example, a direct quote from Joseph Stalin about his motivations regarding the Five-Year Plans would be a primary source. So would a direct quote from a Soviet peasant about living conditions under the Five-Year Plans. A secondary source is a book or a point of view that was not directly involved in the situation. For example, a modern textbook about the Reformation is a secondary source. The book you're reading right now, for that matter, is a secondary source.

The test writers may ask you to identify if a source is a primary source or a secondary source, so make sure you understand the difference. More often, however, they will use sources in the questions and then ask you to make generalizations from the source.

Trends

The test writers will occasionally ask you to infer a trend from a series of events. You should be knowledgeable of the general trends that have occurred in the world since AD 1000. A list of important trends is below.

- Cultural diffusion and interaction have been accelerating since AD 1000. Early on, migrations of people, trade, and the crusades were the main source of interaction. Then came the steam engine. Mercantilism. The telephone. Airplanes. Television. The Internet. And so on.

- A trend toward democracy has emerged since AD 1000. Monarchies and totalitarian governments have given way to free elections. The Magna Carta was a major step. So was the Enlightenment. So was the French Revolution. So was the end of World War II. So were the independence movements. There have been setbacks along the way, but in general, the world appears to be solidly moving toward democracy.

- There is a trend toward westernization and modernization. The Industrial Revolution was, of course, extremely important, but political decision-making has also had a huge impact. Peter the Great, the Meiji Restoration in Japan, Ataturk, and the fall of communism have all been important steps along this path.

- For a long time increased warfare was the trend. However, while our capacity for warfare has increased since World War II, for the first time in history major nations are restraining themselves from using their weapons technology.

- There is a trend toward urbanization. In AD 1000 , the vast majority of people lived on farms. Now, the majority of people live in cities, especially in the industrialized nations. The Industrial Revolution was a major step in this trend. With urbanization, there is a trend away from the traditional value system in many cultures.

Of course, there are many trends throughout history. These are just a few.

If the test writers ask you to identify a trend or to look at a series of events and form a conclusion, keep in mind the general development of world civilizations.

Timelines

The test writers will expect you to be able to put major global events in the correct chronological order. They may present you with a timeline or a list of events. As you review the information in this chapter, make sure you pay attention to events that resulted from and/or led to other events. Make sure that you also understand events that occurred contemporaneously.

A few of the major sequences of events are:

- World War I led to the Treaty of Versailles, which led to an economic depression in Germany, which led to the rise of Hitler, which led to World War II.

- The Scientific Revolution led to the Industrial Revolution, which led to the need for raw materials and markets, which led to colonialism and mercantilism, which led to European cultural diffusion throughout the world, which led to twentieth-century independence movements.

- The Industrial Revolution led to poor working conditions for millions of Europeans, which led to the development of Marxist ideology (communism), which led to the Bolshevik Revolution in Russia, which led to the Cold War.

- The fall of the Roman Empire led to the development of Constantinople as a center for learning, which led to increased interaction with the Middle East and the Islamic Empire, which led to the rise of the Islamic Empire, which led to Islamic expansion into Europe, which led to a European Renaissance. One big circle.

> As you continue to review the content in this chapter, make sure you understand how historical events are interconnected.

Linked historical events go on and on. These are merely a few of the most obvious ones.

Distribution

Distribution questions will ask you how various human or nonhuman characteristics are distributed throughout the globe or within a society. These questions might focus on population, religion, or on resources. To a certain degree, distribution questions are covered in chapter 6 since many focus on geography.

The big distribution issue to focus on for now is that of religion. After reading this chapter, do you understand how the different religions spread around the globe? You should.

> Follow the development of Christianity, Judaism, Islam, Hinduism, and Buddhism as you review the information in this chapter a second time.

Answers to quizzes

Initial practive quiz

1 C	2 G	3 C	4 F	5 A	6 F	7 B	8 F
9 C	10 F	11 A	12 G	13 D	14 J	15 A	

Quick Quiz 1: 1 C	2 F	3 B				
Quick Quiz 2: 1 D	2 H	3 C				
Quick Quiz 3: 1 A	2 F	3 D				
Quick Quiz 4: 1 C	2 G	3 A	4 G			
Quick Quiz 5: 1 C	2 J	3 B	4 F			
Quick Quiz 6: 1 A	2 J	3 C	4 H	5 A	6 F	7 B
Quick Quiz 7: 1 A	2 G	3 A	4 F	5 D		
Quick Quiz 8: 1 C	2 J	3 B	4 J			
Quick Quiz 9: 1 C	2 H	3 A	4 J	5 A		

Chapter 6

Review of Geography since AD 1000

What You Need to Know

First things first: Review chapter 5 before reviewing this chapter. The geography information in this chapter is dependent on historical context. The geography questions will require you to look at historical facts and apply the correct geographical concept to the question. If you understand all of the historical facts in chapter 5, you'll get more from this chapter than if you don't.

Keep in mind that you're not expected to know everything from your tenth-grade geography class. Some of the stuff from your geography class will be tested on the World History and Geography to AD 1000 exam, some will be tested on the World History and Geography from AD 1000 exam, and some of the stuff won't be tested at all. For the purposes of the World History and Geography from AD 1000 exam, it just makes sense to focus on the impact of geography since AD 1000. This chapter is designed to help you focus on the right stuff.

You should have already studied these categories in chapter 5:

1. Late medieval Europe through the Reformation — 8 questions

2. The Age of Discovery — 8 questions

3. Sixteenth–nineteenth centuries: Enlightenment, Absolutism, Reason, and the Industrial Revolution — 12 questions

4. Twentieth-century world conflicts — 10 questions

5. History skills — 7 questions

Now, in this chapter, you'll study two new categories:

6. Geography knowledge and concepts — 12 questions

7. Geography skills — 6 questions

> If geography skills questions bug you, then study geography knowledge and concepts questions to make up for it.

Geography questions ask about the impact of humans on the land or the impact of the land on humans. They also ask about the characteristics of maps and mapping and about the distribution of cultural characteristics throughout the world. In other words, geography questions ask you to apply general historical information to an understanding of geography.

Geography Knowledge and Concepts

There are 12 geography skills and concepts questions on the exam. They test:

- The impact of technological advances

- The impact of natural hazards

- The impact of resources

- Developed and developing countries: The big picture

- Economic interdependence

- Comparative advantage

- Conflict and cooperation among countries

Impact of Technological Advances

Technology makes the world smaller

One of the main features of the study of geography since AD 1000 is humanity's ability to affect it with technological advances. In the beginning, these advances included rudimentary forms of irrigation and levee-building and so forth. In the years following the Industrial Revolution, technological advances became far more sophisticated. The test writers will focus on the development of canals, automobiles, and airplanes, as well as advances in communication and the way in which technology has made the world smaller.

The shortest distance between two points is a straight line

Canals are man-made waterways connecting two natural bodies of water for transport purposes. Before the development of canals, ships were at the mercy of the natural lay of the land. If a huge continent stood between point A and point B, the ship had to go around the continent to travel between the two points. Canals were built at narrow points in the land to create new, more efficient routes. The two most important canals are the Suez Canal and the Panama Canal, which together allow ships to travel around the globe near the population centers rather than having to go around the edges of the continents. Both Egypt and Panama are commercially and militarily strategic countries in large part because of the existence of the canals.

Planes, trains, and automobiles

The invention of the steam engine transformed the landscape because it not only led to the creation of large cities under the factory system but also connected cities through the railroad. Trains transformed the way people thought about distance and geography, and commerce and cultural customs spread accordingly. With the development of automobiles and airplanes, the world would never be the same. World travel is now commonplace, and not just among business leaders and politicians.

Of course, technological advances in transportation have also severely impacted the environment, from the emission of fossil fuels into the atmosphere (leading to pollution and, many believe, global warming) to the transformation of the landscape with the creation of highways, airports, and seemingly endless suburban sprawl. Cities, especially in the United States, are much more spread out than they used to be, in large part because of the automobile. It is now highly uncommon for an American resident to walk to work or to take mass transportation. And since more Americans now live in the suburbs than in the urban core, most Americans have now become *dependent* on the automobile, which is an enormous cultural change from just 70 or 80 years ago.

Technology and communication

Of course, the impact of technology on communication is remarkable. It truly has become a small world after all. You already know how easy it is to communicate with someone hundreds or thousands of miles away. What you need to remember for the test, however, is that this ability is new and still transforming. Just ten years ago, e-mail was used in universities, some businesses, and in the government, but not in the home. Twenty years ago, e-mail didn't exist in its current form—nor did cell phones or fax machines. Fifty years ago, transcontinental phone calls were unheard of, and many Americans were sharing phone lines with neighbors. A hundred years ago? The only way to communicate was via letter or personal visit.

> Phones, e-mail, pagers, and fax machines are just a few ways that you can be reached that weren't readily available 100 years ago.

The world at your doorstep

Of course, radio, movies, television, and the Internet have made everything in the world available. Again, think about the impact of these inventions on attitudes and on world culture. Imagine if you *never* watched television or movies and *never* listened to the radio or surfed the Web. Your life would be so different you probably wouldn't even recognize it.

Clearly, transmitters, satellites, and computers have transformed our lives. What you need to understand is that these things have also transformed business and politics. If you sell something over the Internet to someone in Bangladesh, who collects the taxes? If you write an e-mail to someone in China, can it be censored by the government? If a country doesn't recognize United States copyright laws, can you transfer copyrighted material to that country over the Internet without a penalty? The complications continue to mount, in large part because technology is growing faster than our laws can change.

The digital divide

While the test writers will focus primarily on technology's impact on transportation and communication, technology has impacted everything, from farming methods to health care options to sports and education. However, you need to remember that the impact of technology is felt differently in different parts of the world. Developed countries, like the United States, Canada, Japan, and those in Western Europe, for example, are responsible for much of the technology, and they have made technological advances available to the average person within these countries. Many of the people in the developing world, however, don't even have telephones, much less a computer.

Since technology is growing so quickly, the gap between the technologically connected and the unconnected is growing by the month. This gap is known as the **digital divide**, and it divides not only one country from another but also individuals within countries. It doesn't matter if you grow up in Richmond or in Thailand—if you're connected, then you're part of the digital community, and if you're not, you're not. How all of this plays out in business, politics, distribution of resources, and effectiveness of borders and boundaries is the challenge of your generation.

If you don't think you are technology-dependent, think about what would happen if someone took away your telephone and computer.

The Impact of Natural Hazards

Regardless of technological advances, natural hazards continue to occur. Technology has not yet prevented natural hazards (in fact, some argue that technology has affected our climate so much that it actually *creates* natural hazards), but it has had an impact on a society's ability to deal with natural hazards.

In short, the test writers will want you to understand that developed countries are better able to handle natural disasters than developing countries. Countries like the United States, Germany, and Japan not only have early-warning systems in place to warn citizens of potential disasters, but they also enforce strict building codes to reduce the damage done by fires, tornadoes, and earthquakes. An earthquake of the same magnitude that strikes a city in the United States and a city in Southeast Asia will often do much more damage to the Southeast Asian city.

What's more, developed countries often have the money and resources to quickly rebuild in the event of a devastating natural disaster. In the developing world, rebuilding efforts often take much longer and often involve poorly constructed buildings, once again setting the country up for another disaster.

The Impact of Resources

What to do with this oily goo?

What makes natural material a natural resource? Usefulness. Plain and simple.

If a natural material is useful to mankind, we consider it a resource (like oil, trees, and water). If it is not useful to mankind (like pond scum), we don't consider it a resource. The incredible (and some people think scary) thing about human achievement (or mismanagement) is that most natural materials have become natural resources precisely because humans have figured out ways of using them. Things we now call food were once just natural materials that we figured out we could eat. Stuff we now call sources of energy were once thought of as pools of goo below the Earth's surface.

What's more, some resources have grown in usefulness as humankind has figured out more ways to use them. Water, for example, is not only a resource essential for human consumption, but also a resource because it allows us to make use of other resources (like soil and seeds)

159

and provides us with other kinds of resources (like fish). Water became an even more important resource when people figured out how to use it for transportation, and a more important resource still when people figured out how to use it to generate electric power.

Now you know that natural resources and a stable political system bolster the United States' position of power in the world.

Here's the part that makes life on Earth so complicated: resources are not distributed equally. The United States is jam-packed with resources—land, water, coal, oil, forests, rivers, and so on. It's no coincidence that the United States is also arguably the most powerful economy in the world. Other factors, such as a stable political system, have also impacted the United States' stronghold, but resources have played a huge part.

Don't stop thinking about tomorrow

Some resources are renewable, like soil and water. In other words, soil, if it's treated right and allowed time to replenish, can serve farmers in 100 years just as it serves farmers today. Water changes forms and location, but it is also renewable. Every time water evaporates, it is repurified and returns to the Earth as rain, feeding the farmland and making its way down into a river, where perhaps it generates hydro-electric power.

Other resources such as oil, coal, and natural gas are not renewable. Once you use them, they are gone. Well, actually, it's worse than that. The resource is not *gone*, it's simply changed into byproducts, which are not useful and which in many cases are harmful pollutants. The problem, of course, is that we've structured our world on the consumption of nonrenewable resources. Eventually we'll run out of oil, coal, and natural gas, and we'll have to develop some new sources of energy.

Not all resources come from dirt

Natural resources aren't the only important resources. Human resources and capital resources are important as well.

Human resources simply means people resources. If you want to move furniture, you need a lot of people who are strong. If you want to cure AIDS, you need a lot of people who are smart, highly educated, and experienced in medical research. If you want to plant a garden, you need lots of people who are knowledgeable about different kinds of plants and skilled at growing them. In other words, human resources are people who have the characteristics, knowledge, or talent to do the jobs that need to be done.

Countries with a lot of human resources have strong education systems, strong job-training systems, and lots of incentives for innovation and creativity. These countries have lots of highly skilled, well-paying jobs.

Capital resources simply means money. A country such as Japan may not have a lot of natural resources, but because of the way it has managed its business and education systems, Japan does have a lot of capital and human resources. Since Japan has a lot of capital resources, it can invest in its own future. Conversely, many countries in Africa and South America have natural resources but don't have the capital resources to invest in the infrastructure needed to exploit their resources. They either have to borrow money, wait for foreign companies to invest, or slowly build up capital and human resources one step at a time.

In short, countries with extensive resources (natural, human, and/or capital) and the ability to use those resources to improve the economy and living standards are known as *developed countries*. Countries that are in the process of developing their resources are known as *developing countries*.

Developed and Developing Countries: The Big Picture

Developed and developing countries are often compared. The charts below will help you to review the differences between the two groups. Keep in mind that no country is "totally developed" or "totally not developed." Instead, countries fall along a scale. A country is said to be "more developed than" another country based on a certain indicator. Still, we tend to say that the United States, Canada, Western Europe, Japan, Australia, New Zealand, South Africa, and Israel are part of the developed world, and that most of the rest of the world is developing, but at various degrees.

Economic development indicators

	Developed	Developing
Infrastructure (including roads, airports, communications systems, energy systems, waste disposal, water treatment, energy production, and food distribution systems.)	highly advanced throughout country only in large cities	under-developed or developed
Labor force	highly skilled, well-educated	less-skilled, time-intensive jobs
Gross national product (GNP) Gross domestic product (GDP)	generally high ($20,000+ per capita)	generally low ($5,000 or less per capita)
Population distribution	primarily urban	primarily rural
Educational opportunities	majority with high school diploma; large numbers with college degrees	little education past eighth grade
Availability of resources	extremely high	extremely low

Standard of Living and Quality of Life Indicators

	Developed	Developing
Population growth rate	moderate	extremely high
Population age distribution	fairly even	lots more young than old
Literacy rate	often 90% +	often less than 50%
Life expectancy	close to 80 years	close to 50 or 60 years
Infant mortality	very low	very high
Percent involved in agriculture	very low	very high

Economic Interdependence

Economic interdependence results when the economy of one country depends on goods or services from another country. For example, the industrialized world is dependent on the Middle East for petroleum, without which industry would be halted. Most industrial nations do not have enough petroleum of their own and yet built an economy that requires it. In addition, many countries in the Middle East have structured their own economies around the sale of petroleum. Therefore, if the industrial world finds an alternative fuel, development in the Middle East would be severely slowed.

As countries trade more, their economies and cultures become dependent on the products and ideas received through trade. As such, there has been an increasing amount of activity surrounding trade negotiations between countries and even the establishment of quasi-governing bodies to enforce trade regulations and keep information, products, and services moving. Some countries have become so dependent on the success of partner countries that they have formed economic unions. **The European Union**, for example, allows for free trade among its members while also unifying Europe in economic competition with the United States and Japan. On this side of the Atlantic, **NAFTA** is a trade agreement between Canada, the United States, and Mexico to keep the trade doors open among those three countries.

Comparative Advantage

Comparative advantage occurs when a country has a surplus of a resources, products, or services that the rest of the world envies. Such a surplus gives the country an advantage because it can trade the surplus at a high rate of exchange without having to sacrifice its use among its own people. Highly developed countries have comparative advantages. Countries that are not highly developed are at a comparative disadvantage. Often, they struggle to meet their own needs (they are subsistence economies with few surpluses) and must either borrow from the developed countries or sacrifice their own products in exchange for other products that they also need.

When a country has a comparative advantage, it often leads to job specialization and an increase in human resources. This, in turn, leads to an even stronger comparative advantage, which leads to more job specialization and so on.

Conflict Between and Cooperation Among Countries

Conflict between and cooperation among countries are largely questions of geography because they involve international borders, distribution of resources, and cultural ties and differences. The test writers will want you to understand some basic concepts involving international conflict and cooperation and a few examples of each concept.

Imperialism and colonization

Imperialism or **colonization** is a nation's practice of invading another nation by taking control of its government and economy. The new territory is referred to as a colony of the mother country. During the eighteenth, nineteenth, and early twentieth centuries, European colonialism led to its domination of almost the entire globe. Colonialism was an important method by which colonizing nations gained access to raw materials and markets during the Industrial Revolution. It has been a major cause of European cultural diffusion throughout the globe and has, to this day, resulted in political tensions between the developed and the developing world. The developed world sometimes argues that through colonization it brought increased technology and opportunities to the colonies. The developing world argues that colonization stripped its colonies of resources and lined the pockets of the colonial power. Much of the world has won its independence from the grasp of colonialism, and few people argue in favor of that kind of system, but most current world political conflict and cooperation is reminiscent of imperialism.

> Even after power has been returned to the indigenous population, a former colony can be affected by colonization, especially if it has already been robbed of its natural resources.

Partition

A **partition** occurs when a territory is split between two or more groups to become separate territories, or split between two existing territories and absorbed by those territories. For example, in the late 1930s, Germany and the Soviet Union agreed to a nonaggression pact. Even though they didn't like each other, they agreed they wouldn't attack each other so that each could focus on its other enemies. As part of that agreement, they agreed to a partition of Poland (without Poland's consent, of course) in which each country would acquire part of a defeated Poland.

Perhaps the best example of a partition, though, is in the case of the Indian subcontinent. After India was colonized by Britain, the entire subcontinent was considered British India. It included large populations of both Hindus and Muslims. As India moved toward independence, however, it was clear that some people wanted a united India and others wanted the subcontinent partitioned into two different countries, one Muslim and one Hindu. Britain helped broker a partition agreement, and after independence, the sub-continent was split into three parts: India (mostly Hindu) in the middle, Pakistan to the northwest, and East Pakistan (now Bangladesh) in the southeast.

Another good example of a partition is that of Ireland, with Northern Island partitioned as part of the United Kingdom. Cyprus is yet another example, split in half between the Greeks and the Turks. As shown in all of these cases, however, a partition does not necessarily result in an end of conflict.

Nationalism

Nationalism is loyalty to a nation based on shared customs, language, and history. Nationalism has led to independence movements (such as those in Africa against colonial powers) as well as the combination of smaller nations into a larger one (the reunification of Germany, for example). In its extreme, nationalism leads to a feeling of superiority that can be dangerous (Hitler's Germany, for example).

Nationalism is a relatively new concept. Before the nineteenth century, many people identified themselves with their particular social class, their religion, or their particular location. But since then, nationalism has swept the globe. In Africa, national governments have spent decades trying to instill nationalistic feelings in the hearts of their people because many people are loyal to their tribes, not their nations. Meanwhile, the European Union is eroding nationalism by trying to get people to think of themselves as Europeans, not as French or German, for example. In any case, nationalism continues to have an enormous impact on how large groups of people choose to structure their societies, borders, alliances, and conflicts.

Treaties and alliances

Treaties are agreements between countries for a mutual benefit. The United States and Russia, for example, often try to negotiate arms-control treaties under which both countries promise to reduce their nuclear weapons arsenals. Or two countries may agree to a nonaggression treaty under which they promise not to attach each other.

Alliances are like treaties and often stem directly from treaties. Alliances typically go a bit deeper than the mere words of a treaty. Instead, they usually involve a substantial relationship between countries, sometimes for a specific purpose and sometimes for a very general purpose. NATO and the Warsaw Pact are two important recent alliances. Both were formed in response to political shifts after World War II. NATO was (and remains) a military alliance between Western European nations and the United States for the purpose of containing Soviet communist expansion. The Warsaw Pact was a military alliance between the Soviet Union and its Eastern European satellites for the purpose of achieving common objectives.

When a country breaks a treaty or an alliance, the results can be momentous. When Germany broke its nonaggression treaty with the Soviet Union near the beginning of World War II, the Soviet Union entered into an alliance with Britain and the United States for the sole purpose of defeating Germany. After World War II, when the Soviet Union didn't live up to its treaty promises regarding free elections in Eastern Europe, the Soviet Union and the United States created European alliances against each other (NATO and the Warsaw Pact).

> Treaties and alliances rarely last forever. They shift according to the political and emotional climate of the times.

The United Nations

The **United Nations**, created after World War II to discuss and solve problems in the hopes of avoiding another catastrophic war, includes almost every country in the world. It has no authority over its members but is nevertheless an increasingly important and powerful body. It has many units and independent agencies, such as the World Bank and the World Health Organization. It collectively addresses world issues like the spread of disease, economic development, deforestation, political instability, human rights violations, ethnic violence, and so on. Its military forces were instrumental in the Persian Gulf War, and its peacekeeping forces have been utilized throughout the globe.

In short, the United Nations is one way in which nations work to cooperate with each other or to pressure nations that aren't cooperating with each other into doing so.

Geography Skills

The exam includes six geography skills questions. The interesting thing about these geography skills questions is that they are actually much more knowledge-based than skills-based. Essentially, these six questions will ask you to identify rivers, mountain ranges, deserts, countries, climatic patterns, and other physical features. The people in charge of the SOLs have released a list of the geography features they consider to be the most important, and we've organized information according to the following rules. As you review each world region listed below, consult a world map or globe so that you are clear about the location of each geographic feature relative to other geographic features.

General Rules of Geography

Regardless of the region of the world, you need to keep in mind a few general rules about geographic features. If you know these rules, you'll be able to eliminate incorrect answer choices to geography skills questions very easily, even if you don't remember the details of the particular geographic feature in the question.

Rule #1: Deserts make people feel deserted

If you aren't certain of a correct answer, choose the answer choice that uses general language and makes sense *thematically*, regardless of the specific example in question.

A desert, as you know, is a dry place with little or no vegetation. Deserts are significant because they serve as regional barriers against transportation and communication, and therefore block cultural diffusion. Historically, deserts separate different cultures. In other words, the society on one side of the desert was often markedly different from the society on the other side. The test writers' favorite desert is the Sahara Desert in Africa, which separates the North

African cultures from the rest of Africa such that the rest of Africa is simply called Sub-Saharan Africa. However, regardless of the desert in question, choose the answer choice that suggests that the desert is a barrier that leads to isolation.

Rule #2: Rain forests are good for the environment but bad for traveling

Rain forests are one of the three big barriers (mountains and deserts are the other two). Rain forests are dense areas of vegetation in a tropical, rainy climate. Hard to traverse, rain forests are generally home to small groups or tribes that have little contact with the outside world. Rain forests are also home to a rich diversity of wildlife, many species of which are not found in other climates. The most significant rain forest is the Amazon Rain Forest in South America.

Rule #3: Mountains get in the way

Since mountains are essentially giant walls of rock and dirt, they are huge barriers to transportation, communication, and cultural diffusion. Mountains isolate, but they also protect. Of course, in today's world, we have the capacity to simply fly right over them or blast a tunnel right through them, but even today, it's difficult to get entire groups of people to the other side. What's more, villages in the mountains are often quite remote. After all, it's pretty difficult to land an airplane on a rocky cliff! So whether the test writers ask about the Andes, the Himalayas, or some other mountain range, focus on answer choices that refer to mountains as obstacles.

Rule #4: Passes make life easier

Mountain passes are like hall passes: they allow people to travel where otherwise they might not have been able to travel. Passes are narrow breaks in a mountain range that allow people to travel through mountains rather than over them. Where passes don't exist or where they are hard to find, civilizations on either side of the mountains are isolated from each other. But where passes exist, people can travel from one side of the range to the other. What's more, everything can pass through a pass, including bad stuff like an enemy army! Khyber Pass, which cuts through the Hindu Kush Mountains and connects India to mainland Asia, is probably the most famous pass of all.

Rule #5: Nothing rivals a river

The test writers love rivers. Rivers were the lifeblood of early civilizations. They provided water for crops and drinking, transportation, and methods of communication. Since early civilizations grew along rivers, rivers became so central to the cultures that they became intertwined with folklore and religion. The first civilizations were all located in river valleys (the Nile, Tigris-Euphrates, Indus, and Huang Ho). Even if the test writers ask about a river valley that you don't know much about, stick with answer choices that are consistent with all river valleys.

Rule #6: Islands cause separation

The locations of island countries have a significant impact on the development of those countries, especially before the invention of the airplane, telephone, television, and the Internet. In short, island countries that are comprised of one main island or a few islands (like Great Britain and Japan) have used this to their advantage, while island countries that are comprised of hundreds or thousands of islands (like Indonesia) have had a hard time establishing a national identity. In Japan's case, the island location curtailed cultural diffusion to the point where, even today, Japan is one of the most homogeneous nations in the world. In Britain's case, the island location led to the development of a strong navy, which led to Britain's rise as a commercial and military power, which further led to the development of an empire. In Indonesia's case, the existence of so many islands in one country led to difficulties uniting the country since each island exists on its own.

Rule #7: Irregular coastlines are a ship's best friend

Many conflicts in history have revolved around access to a coastline, and therefore, trade.

A jagged, irregular coastline typically has many ports and protected bays. Therefore, irregular coastlines are well suited for manufacturing and transportation centers (cities). The irregular coastline of Western Europe, for example, has numerous safe ports for large ships, so it shouldn't be surprising that Western Europe developed a large-scale shipping industry much earlier than Africa and Asia, which both have long stretches of regular (smooth) coastline. In short, the test writers like to focus on the idea that an irregular coastline aids a country's naval capacity.

North America

The major countries in North America are the United States and Canada. Mexico is also part of North America, but the test writers will include it with Latin America (Central and South America) because of the cultural similarities. Both the United States and Canada have very high per-capita Gross National Products(GNPs); are economically developed; have enormous natural, capital, and human resources; and occupy large land masses.

On a map, you should be able to identify the following rivers and bodies of water that affect North America: the Atlantic, Pacific, and Arctic Oceans; the Caribbean Sea and Gulf of Mexico; the Great Lakes; the Missouri/Mississippi, St. Lawrence, Colorado, Columbia, and Rio Grande River systems. You should also be able to identify the following mountain ranges: Appalachian, Rocky, Sierra Nevada, and Cascade.

You won't need to know details about the waterways and mountain ranges listed above. It is more important that you understand that North America is extraordinarily resource-rich, from farmlands to coal deposits to forests. In addition, you should understand that because North America is relatively isolated from the rest of the world (separated by two oceans), it has been able to expand peacefully without significant fear of invasions from neighboring enemies.

Latin America

Latin America includes Mexico, Central America, South America, and Caribbean countries such as Cuba and Haiti. These countries are bound together by a common culture rooted in Western European imperialism, Catholicism, and the Spanish language (keep in mind that Brazil, the largest nation in Latin America, was colonized by Portugal, not Spain, and therefore, Brazilians speak Portuguese, not Spanish). Most of Latin America is considered part of the developing world, though Brazil and Argentina are swiftly developing and have emerging middle classes.

Latin America's most significant river is the Amazon River system. It flows primarily through northern Brazil, where it empties into the Atlantic Ocean. Much of its watershed consists of the Amazon Rain Forest, which is sparsely populated and home to an incredible array of wildlife. The Amazon River and Rain Forest are major geographical barriers in the region. Still, development in the region has consistently increased over the past century and threatens to damage the environment significantly as loggers chop down huge swaths of forest. What's more, pollution from both urbanization and modern farming methods threaten the river water.

The most significant physical feature in Latin America is the **Andes Mountain Range** near the west coast of South America. The border between Chile and Peru is formed by much of the range. Like mountain ranges in other parts of the world, these mountains have served as a barrier to development and communication in western South America.

Finally, you should also be familiar with the **Panama Canal**, which was built on the Panama isthmus at the close of the nineteenth century. The canal provides a shortcut for boats traveling between the Atlantic and Pacific Oceans. One of the most heavily trafficked waterways in the world, the canal has created many jobs for the people of Panama. However, the canal also has made the region strategically important to the industrialized world, which has led to a lot of foreign involvement in Panama. The United States, which controlled the canal throughout the twentieth century, has invaded Panama on several occasions to ensure that its government and policies are favorable to U.S. interests in the region.

Because the U.S. controls the Panama Canal, it has great influence over not only Panama, but also trade worldwide.

Western Europe

Western Europe is highly developed economically, militarily, and politically. The largest countries in terms of land area are France, Sweden, and Spain. In terms of population, the largest countries are Germany, United Kingdom, France, and Italy. In terms of per-capita GNP, the most prosperous countries are Switzerland, Luxembourg, and Sweden, although all of the Western European countries have relatively high GNPs.

Important rivers and bodies of water include the Atlantic Ocean; the North, Mediterranean, Aegean, and Adriatic Seas; the Strait of Gibraltar; and the Rhine, Seine, and Danube Rivers. Important mountain ranges include the Alps, the Pyrenees, and the Apennines.

You need to know a few big-picture concepts as well.

First, Western Europe is resource-rich, especially in coal. The presence of coal allowed it to develop swiftly during the Industrial Revolution.

Second, Western Europe has an irregular coastline with plenty of deep, protected harbors. This helped to encourage a booming shipping industry, which led to Europe's unparalleled success at exploring and colonizing much of the rest of the world.

Third, Western Europe is densely populated and extraordinarily diverse. Although perhaps to Asian or African eyes Western Europeans seem remarkably similar, throughout much of the past thousand years, Western Europeans have seen great differences in terms of language, customs, and variations of the dominant religion, Christianity. Since Western Europe is so densely populated with so many small nations sitting side-by-side, it has been a hotbed of clashes and wars for centuries. Because Europe industrialized quickly and always used the latest technologies in its warfare, these wars (World Wars I and II in particular, but there have been hundreds of wars throughout the centuries) have devastated the people and the landscape.

Fourth, the **Gulf Stream and North Atlantic Drift** (warm ocean currents) help make Western Europe a great place to live. Largely driven by the winds, the currents keep the water moving, thereby stabilizing the climate on much of the earth. Without them, the tropical regions would become hotter and hotter while the polar regions grew colder and colder. The Gulf Stream and North Atlantic Drift help to moderate the temperatures of Great Britain, Iceland, and Norway in particular, but much of Western Europe as well. Great Britain, for example, enjoys temperatures more similar to those of New York City than those of Moscow, even though nearly half of the nation is as far north as Moscow. Undoubtedly, these moderate temperatures have influenced Western Europe's agricultural and commercial development.

Eastern Europe and Northern Eurasia (Russia)

This region of the world includes much of the area occupied by the former Soviet Union and its satellite nations. The largest countries in terms of size are Russia, Kazakhstan, and the Ukraine, and the largest in terms of population are Russia, the Ukraine, and Uzbekistan. The countries with the highest per-capita GNPs are Estonia, Latvia, Russia, and Belarus.

Significant bodies of water that you might be asked to identify include the Pacific and Arctic Oceans; the Black and Caspian Seas; the Bering Strait; and the Lena, Ob, Amur, and Volga Rivers. Significant geographic features include the Ural and Caucasus Mountains, and the Great European Plain.

A few big-picture concepts merit discussion.

First, Russia's geographic location has influenced its history significantly over the past 1,000 years. During most of Russia's history, the country lacked a warm-water port. This stifled Russia's military and economic ambitions by partially isolating it from the rest of the world even as the rest of Europe was off exploring and conquering the globe. This single fact shaped Russia's military policies for decades, particularly during the reigns of Peter the Great and Catherine the Great. Peter conquered lands to the west of Russia and built a new capital at St. Petersburg, Russia's "window to the West." Catherine conquered lands to the south and established warm-water ports on the Black Sea. Even to this day, geography significantly impacts Russia's decisions. With the break-up of the Soviet Union, Russia once again finds itself land-locked and is impatient with the prospect of independence movements along its southern border.

Second, Poland's location on the **Great European Plain** (a large, flat plain in central Europe) has affected its history significantly. Poland has benefited from the plain, which is not only agriculturally rich, but also rich in resources such as coal, copper, natural gas, and sulfur. Politically, however, Poland has been cursed by the land, which does not offer many natural barriers against invasion, especially from the west, east, and southeast. Over the past few centuries, Poland has been carved into pieces as a result of invasions from Prussia, Austria, Russia, Germany, and the Soviet Union. It now relies on its alliances with the West, most notably with NATO, to give it the protection that its topography and location cannot.

Third, the Bosphorus, a narrow, twenty-five-mile strait that connects the Black Sea to the Sea of Marmara (which connects to the Mediterranean Sea) has played a huge role in political developments in Eastern Europe and northern Eurasia. Because it links the interior waterways of Eastern Europe and Russia with the rest of the world, those who control the strait can (and have) enormously affect Eastern Europe's and western Asia's ability to maneuver commercial and military fleets.

> Though it does not hold as much territory as it used to, Russia's size still makes it a major influence in Europe and Asia.

Middle East and North Africa

The Middle East includes the countries between the Persian Gulf and the Mediterranean Sea, plus Turkey and North Africa (the Sahara and Mediterranean regions). The largest countries in terms of area are Algeria, Saudi Arabia, and Libya, and the largest in terms of population are Iran, Turkey, and Egypt. The countries with the highest per-capita GNP are Kuwait, United Arab Emirates, Qatar, and Israel.

You should be able to locate the Mediterranean, Red, and Arabian Seas; the Persian Gulf; the Nile and Tigris-Euphrates Rivers; and the Strait of Hormuz and Boshporus on a map. In addition, you should be familiar with the Atlas, Taurus, and Zagros Mountains.

Throughout much of the history of civilization, this region was the crossroads of the world. In other words, it's been home to a lot of cultural diffusion. This region is mostly desert, so early cultural diffusion was aided by nomadic trading, but recently the oil-rich economies of the Middle East have made great improvements in transportation networks. The desert is broken up by the Fertile Crescent, which includes the Tigris-Euphrates River valley and the Nile River valley. These river valleys were home to two of the world's first civilizations and still contain huge population centers.

You should understand several concepts regarding the impact of geography on the Middle East since AD 1000.

First, the Byzantine Empire, Ottoman Empire, and now the modern republic of Turkey have all controlled the Bosphorus, which is the narrow strait that connects the Mediterranean Sea to the interior waterways of Eastern Europe and Russia. Therefore, the region around the Bosphorus has always been strategically important. One of the reasons Turkey is part of NATO even though it is culturally allied with the Middle East is because the West recognizes the importance of the Bosphorus to regional security.

Second, the Fertile Crescent remains the most agriculturally productive and livable area in the Middle East. In recent centuries, the people of the Middle East have significantly altered their environment, especially along the main waterways. New systems of irrigation have resulted in increased agricultural production. Along the Nile, a series of dams have been built, generating electric power and altering the traditional flood cycle along the river.

Third, since the region contains so much petroleum, and since petroleum is of essential importance to the functioning of the modern world, the Middle East is one of the most important areas of the world. Nearly two-thirds of the world's petroleum reserves are located in the Persian Gulf region. The presence of petroleum has greatly enhanced the economies of the Persian Gulf nations, and as a result, the quality of life of many Middle

Easterners has improved. The regional governments have spent some of the oil money on educational programs, transportation and communication networks, and scores of other public projects. Still, the presence of petroleum has led to political instability. In the early 1990s, for example, Iraq invaded Kuwait to gain control of its petroleum, and the United Nations, led by the industrialized nations of the world, reacted by defending Kuwait and militarizing the region.

Fourth, the **Sahara Desert**, which stretches in a 1,000-mile-wide path from western Sahara and Mauritania on the Atlantic Ocean to Egypt and northern Sudan on the Red Sea, has virtually swallowed up much of North Africa (and it continues to expand). Traditionally, the desert served as a barrier between the Mediterranean, Middle Eastern, and Egyptian Empires and those of sub-Saharan Africa, but the desert was not entirely impenetrable. Mali and Ghana, for example, sent traders on caravans to exchange ideas and resources with peoples north and east of the desert. Yet, overall, the desert has been a huge barrier to development and to contact with the rest of Africa. It has led to the development of two unique Africas—a Sahara region that is mostly Muslim and Arabic-speaking and allied with the Middle East, and a sub-Saharan region that developed independently and is largely Christian, animist, and affecteded mostly by a combination of traditional African culture and European imperialist culture.

Fifth, because Europe, Asia, and Africa are all connected in one landmass, the building of the **Suez Canal** where Africa meets Asia changed shipping patterns and the fate of the region forever. Prior to the building of the canal, if a ship wanted to travel from Britain to India, it had to go all the way around the tip of Africa and then northeast again. However, a small strip of land in western Egypt changed all that. It was here that the French and British, who colonized the region and had enormous commercial investments at stake, logically built the Suez Canal, which opened in 1869. Travel time between Europe and Asia was reduced dramatically, and Egypt became strategically important. After Egypt achieved independence from Britain in the 1920s and then nationalized the canal in 1956, its economy was boosted by the fees it charged for use of the canal.

Sixth, certain regions of the Middle East are so important to the history, religion, and culture of some Middle Eastern groups that the territorial disputes over these regions have caused nothing but constant bitterness and warfare. For example, Palestine, a region on the eastern shore of the Mediterranean Sea, is land that Jews hold as sacred because they believe it was promised to them by God. After the Diaspora, the region converted to Islam, and Palestine became home to Muslims. Muslims that reside in the area are known as Palestinians, and Palestinians believe that Palestine rightfully belongs to them. But as a result of the Zionist movement before and during World War II, modern Israel was established in Palestine as a Jewish homeland in 1948. Ever since, Israel and Palestine have battled for control of this region. Of particular concern are several areas: the West Bank (a region on the west bank of the Jordan river to the east of Israel that both Israel and the Palestinians claim), the Gaza Strip (a small band along the Mediterranean between Egypt and Israel), and the Golan Heights (a small region between Israel and Syria). Disputes over these lands have led to numerous skirmishes and peace settlements since the 1950s. Currently, Israel is willing to release some its claims on land in exchange for security and peace in the territories it solidly occupies, but true peace has not yet been achieved.

> The discovery of oil in the Middle East has made many Arab nations prosperous while adding to the already existing conflicts in the region.

Sub-Saharan Africa

Africa includes three vastly different regions: The Sahara region in North Africa is culturally part of the Middle East (and is discussed above); sub-Saharan Africa is geographically and culturally distinct from the Sahara region; and the Republic of South Africa at the southern tip of the continent has a history unique from the rest of the continent and is loaded with natural resources, such as diamonds and gold.

In modern-day sub-Saharan Africa, the largest countries in terms of land are the Congo (formerly Zaire), Sudan, Chad, Mozambique, and Madagascar. In terms of population, the largest countries are Nigeria, Ethiopia, Congo, Tanzania, and Kenya. Finally, in terms of per-capita GNP, the richest nations are South Africa, Gabon, and Botswana.

Look at a map of Africa and follow along. In the middle of the continent is the tropical rain forest. Above and below the rain forest is the savanna. Then, above and below the savanna is the steppe. So it's kind of like a dart board, with the rain forest as the bull's eye surrounded by two rings of very different topography. The tropical rain forests are hot, rainy, and dense, and even today are really hard to get through. The test writers might test the idea that rain forests were very hard for the European colonists to settle. The savannas are also pretty unwelcoming places because they are hot in the summer and dry in the winter, with hardly any trees. They're home to a whole lot of wildlife and are decent places for farming, though irrigation in the region would help a lot. Finally, the dry steppes are just that—dry!

The test-writers may include a few concepts about the geography of sub-Saharan Africa.

First, the ancient West African kingdoms thrived, but were relatively isolated from the rest of the world. They had to send caravans of traders across the Sahara to interact with Mesopotamia and beyond.

Second, the coastline is very smooth, so there aren't very many natural harbors for the development of coastal cities and naval centers. This means that sub-Saharan Africa remained relatively isolated from much of the rest of the world for a longer period of time than other remote regions of the world. (The exception is the southern tip of the continent, which was colonized by Europeans much earlier than the rest of the continent and which served as a stopping-off point for European ships on their way to India.)

Third, when the continent was colonized by the European powers during the nineteenth and twentieth centuries, the Europeans carved Africa into pieces along boundaries that served their own purposes, not along boundaries that reflected the African tribes themselves. The consequence was devastating for long-term stability in Africa. Enemy tribes often were grouped together in the same colony, while other tribes were divided between two or more colonies. To this day, conflict in many modern independent African countries stems from the colonial decision to disregard the culture of the African natives. Rival tribes have fought each other in Burundi, Ghana, and Angola, just to name a few countries.

Fourth, devastating droughts, especially in East Africa, have not only resulted in millions of deaths but have also fueled political and tribal hostilities. Even when international aid in the form of money and food is sent to these nations, the aid sometimes doesn't reach the people who need it most. Instead, as was the case in the 1990s in Somalia, the military leaders of warring tribes fail to deliver the food to the people, steal it for themselves, and block access by international workers to starving citizens who are at the mercy of the military leaders.

Fifth, you need to understand the significance of the **Great Rift Valley**, which runs down the spine of central Africa, separating the great forests of the Congo Basin and Zambezi River from the East African nations of Mozambique, Tanzania, Kenya, and Uganda. The significance of this valley is that it acts like a funnel, driving migrations of people in East Africa in a north-south direction rather than in an east-west direction. This has meant that many of the East African nations have constantly had to deal with tribal warfare in other East African nations, even if the warfare didn't start in their own nations.

> Because Africa was colonized with no regard to tribal boundries, conflicts continue to erupt in African nations

South Asia

South Asia is dominated by the Indian subcontinent and includes India, Pakistan, Afghanistan, Nepal, Bhutan, Sri Lanka, and Bangladesh. This is a unique place in large part because of the Himalayas, which separate the subcontinent from the rest of Asia. South Asia is packed with people. India's land area is half that of the United States and yet its population is more than three times as great. In fact, India is the second most populated country on Earth! In terms of per-capita GNP, the South Asian nations are very poor. Even though the economies of the region are growing rapidly, the population is growing quickly as well, so economic benefit is spread out among an ever-increasing population.

The test writers may ask you to identify the Indian Ocean, the Bay of Bengal, the Ganges and Indus Rivers, and the Himalaya Mountain Range.

It is vital that you understand a few big-picture concepts.

First, rivers play an extremely important part in Hinduism, and therefore in India. The Ganges River is the most sacred river in the Hindu religion.

Second, the **Himalayas**—the highest mountain range in the world—have played an important part in the development of the Indian subcontinent, which lies south of the mountain range. Like most mountain ranges, the Himalayas protect people on both sides of the range and isolate them. The mountains stretch from just northeast of Burma all the way to the northeast corner of Pakistan—in other words, they create a wall between the subcontinent and central Asia. This "wall" has had a tremendous impact on the climate, which in turn has had an impact on the economic and cultural development of India. The Himalayas prevent heavy rains, brought to the region by monsoon winds, from crossing into central Asia, thus allowing the clouds to empty over the subcontinent where the extra rainfall benefits the rice paddies. In addition, the Himalayas protect the subcontinent from cold blasts of air that sweep across northern and central Asia. Together, these two aspects provide an ideal environment for the growth of a civilization.

> The Himalayas are a great example of a landmass that affects local peoples and is affected by local peoples.

East Asia and the Pacific

This region of the world includes Mongolia, China, Japan, Taiwan, Philippines, Indonesia, Malaysia, Thailand, Cambodia, Burma, Laos, Vietnam, and North and South Korea. The most populated country and largest country in terms of land area is China, with over 1 billion people densely clustered in the east along the rivers. Several of the areas in this region of the world are unique from one another, so we've split them apart below.

China

China is the world's most populated nation, and the world's third largest geographic territory. The country was relatively isolated from the rest of the world for thousands of years, in large part because of mountains to the southwest, desert to the northwest, the Pacific Ocean to the east, and the jungles to the south. The **Silk Road**, however, was one way in which China interacted with Asia and Europe. Still, it wasn't until the nineteenth century that China was exposed to the rest of the world on a large scale.

As for China's rivers, they generally start in the west and empty into the east. The Huang Ho River valley is home to China's first civilization; the Yangtze River is the center of modern China's industrial base.

Japan

Geography has had a significant impact on the development of Japan. There are two important aspects to consider: the first is the impact of Japan's island location, and the second is the fact that Japan lacks natural resources.

Japan is a nation of islands off the coast of mainland Asia. The water surrounding the nation has provided it with security from military invasions and cultural diffusion. The result has been the development of a society with a very strong national identity and sense of purpose, although its isolation has led to an ethnocentric attitude. Nevertheless, the fact that Japan is an island nation has generally had a positive impact on its culture. It has experienced longer periods of peace than the vast majority of nations of the world and has developed a fishing industry that is second to none. Fishing is a very cost-effective way to feed a growing population since the oceans provide the fish for free, and Japan's mountainous topography permits only limited farming.

Japan is also a small, mountainous island nation with relatively few natural resources. At the same time, its population has been growing dramatically for centuries. Japan has overcome its lack of resources by importing raw materials from other countries. Since the Meiji Restoration in the late nineteenth century, Japan has been a major player in the world economy by effectively managing its imports and exports. In the early part of the twentieth century, Japan invaded other parts of Asia to gain inexpensive access to raw materials it needed. The Japanese followed the trend of mercantilism established by European imperialists centuries before. They robbed their colonies of resources, shipped them back to Japan where they were made into finished products, and then sent those finished products abroad (as well as back to the colonies), where they were sold for a profit. Since the end of World War II,

It's easy to forget that Japan is an island, and a small one at that. However, Japan has some problems as a result of being so geographically isolated.

Japan has continued to manage its imports and exports well, despite the fact that it no longer has colonial holdings. It still imports most of its raw materials from other countries, but it turns those relatively inexpensive raw materials into much more expensive high-tech finished products. Combined with protectionist trade barriers against other nations, Japan has consistently managed to overcome its geographic barriers by making a profit from other nations' natural resources.

Southeast Asia

One huge geographic concept dominates any discussion of Southeast Asia: the monsoons. Southeast Asia, which includes Myanmar, Thailand, Cambodia, Laos, Vietnam, Singapore, Malaysia, Indonesia, and the Philippines, is really made up of two kinds of countries—the mainland countries (the first five in the list) and the island countries (the last three). The island countries, especially Indonesia, are comprised of so many tiny islands that unity is difficult to maintain. This region of the world is replete with seaports and an abundance of hardwoods, rubber, and spices, all of which are sources of trade with the rest of the world. The Mekong River valley is also a major area of rice production.

Australia and New Zealand

It's unlikely that there will be any questions on the exam about Australia or New Zealand, but you should be able to identify their locations on a world map. Note their relation to the Indian and Pacific Oceans and to the Coral and Tasman Seas. Understand that these two countries were colonized by Britain at the height of European world colonization, but that they became highly populated, and controlled, by European immigrants. To this day, Australia and New Zealand are English-speaking members of the Western world.

Chapter 7

Practice Test 1: World History to AD 1000 and Geography

What to Expect

The following practice exam contains 71 questions, which is the number of questions that is scored on the real End-of-Course World History to 1000 AD and World Geography exam (61), plus ten field-test items. *Remember:* The test-writers of the real exam will include ten field-test questions that will be scattered throughout the regularly scored problems. You won't know which 61 out of the 71 questions on the real exam will count toward your score, so always answer every question. Good luck!

Read each question or statement carefully and then choose the letter of the answer choice that best answers the question or completes the statement.

1 Early groups of people that followed herds of animals in order to survive are known as —

A monotheists

B nomads

C ziggurats

D Mesopotamians

2 Merchants and artisans who fled to the cities during the Medieval period discovered that they had exchanged the security of the feudal manor for all following benefits except —

F freedom from feudal taxes

G rights of self-government

H elimination of wars and conflicts

J expanded trade opportunities

3 Civilizations that developed along the Nile, Tigris-Euphrates, and Huang Ho Rivers all had the common advantage of —

A religious beliefs based on monotheism

B urban communities built by iron and steel tools

C government based on male suffrage

D mild climate, fertile soil, and natural waterways

4 Muhammad's significance to Islam equals that of —

F Siddhartha to Buddhism

G John the Baptist to Christianity

H St. Francis Xavier to the pope

J Moses to the Hebrews

5 "If a seignior (noble) has knocked out the tooth of a seignior of his own rank, they shall knock out his tooth. But if he has knocked out a commoner's tooth, he shall pay one-third mina of silver."

—Code of Hammurabi

Which aspect of Babylonian society does this portion of Hammurabi's code of law reflect?

A All men were equal under the law.

B Fines were preferable to corporal punishment.

C Divisions existed between social classes.

D Violence was always punished with violence.

6 One way in which the civilizations of the Sumerians, the Phoenicians, and the Maya were similar is that each —

F developed extensive writing systems

G emphasized equality in education

H established monotheistic religions

J encouraged democratic governments

7 India's earliest civilizations were located in or around —

   A mountainous areas

   B river valleys

   C deserts

   D steppes

8 Read the statement below.

> As civilization moved to agrarian-based communities, religion beame increasingly more focused on fertility.

This statement is probably true because the supporting evidence includes all of the following except —

   F Environmental gods were placated to ensure smooth flow of seasons.

   G Prayers and sacrifices to the deities protected crops from climatic disasters.

   H Villages maintained clay models of shrines as charms against the fury of nature.

   J Pictures of successful hunts were kept as omens to guide future hunts.

9 The structure of many modern states is very similar to that of the medieval church in all of the following ways except that the church held the power to —

   A determine tax levies

   B administer justice

   C enact policies and laws

   D excommunicate members

10 The region of the world around and between the Tigris and Euphrates Rivers is known as —

   F Shang China

   G The Mediterranean world

   H Mesopotamia

   J Indus Valley civilization

11 The Agricultural Revolution had an impact on geography because —

   A it allowed people to stay in one place and build towns

   B it resulted in the movement of people into mountainous areas

   C it led to the decline of cities

   D it eliminated political boundaries

12 All citizens in ancient Athens had the right to attend the Assembly, where they met to hold open discussions and cast votes. This situation is an example of —

   F direct democracy

   G totalitarianism

   H parliamentary democracy

   J absolutism

13 Which of the following statements is NOT true of the people of ancient Greece?

   A They established a monetary system.

   B They traded with other cultures by sea routes.

   C They were the least advanced militarily.

   D They were polytheistic.

14 In Greek city-states, "citizens" were —

   F free women with no political rights

   G men and women who were not slaves

   H adult males, often engaged in business and commerce

   J men, women, and children living within the boundaries of the city-state

15 All of the following are examples of Roman influence in contemporary society except the —

   A council and magistrate levels of city government

   B codes of civil law for administration of justice

   C Justinian Code outlining the rights of people

   D rehabilitation services for repeat offenders

16 The two most significant Greek city-states were —

   F Athens and Rome

   G Athens and Carthage

   H Sparta and Athens

   J Sparta and Carthage

17 After the Peloponnesian War, the Macedonians took control of Greece and spread Greek culture throughout much of the known world under the leadership of —

   A Alexander the Great

   B Julius Caesar

   C Hannibal

   D Pericles

18 A major contribution of the Roman Empire to Western society was the development of —

   F gunpowder

   G revolutionary socialism

   H monotheism

   J an effective legal system

19 Before the rise of Julius Caesar, the government in ancient Rome was best described as a —

   A direct democracy

   B republic

   C monarchy

   D totalitarian regime

20 The teachings of Confucius encouraged people to —

   F put their own interests first

   G reject government authority

   H prepare for reincarnation

   J follow a code of moral conduct

21 The period of peace and prosperity in Rome under the rule of Augustus is known as —

   A the golden age of Pericles

   B the *Pax Romana*

   C the Edict of Milan

   D the Reformation

22 A major reason for the decline of the Roman Empire was —

    F a series of military defeats in Africa

    G political corruption and an unstable government

    H the abolition of slavery throughout the empire

    J continued acceptance of traditional religions

23 After the fall of Rome, the eastern portion of the Roman Empire became known as the —

    A Persian Empire

    B Byzantine Empire

    C Mongol Empire

    D Gupta Empire

24 The ancient Roman territory known as Gaul is now —

    F England

    G France

    H Germany

    J Russia

25 Which of the following is a primary source?

    A An article about the Peloponnesian War by a famous twentieth-century historian

    B An eyewitness account of the Punic Wars from a citizen of Carthage

    C A television account of the Roman Empire, from rise to fall

    D Correspondence between two medieval monks regarding Aristotle's philosophy

Read the following passage, then answer questions 26 and 27.

> We are called a democracy because the administration is in the hands of many and not of a few. Distinguished citizens are those in public service, not the most privileged. Whereas the Spartans from early youth are undergoing laborious exercises to make them brave...We are lovers of the beautiful. We cultivate the mind without loss of manliness. We gladly toil on our city's behalf.

26 According to this statement, Pericles probably believes that the essence of a city's survival is —

    F maintaining strong civil defense

    G rewarding privileged citizens

    H ensuring citizen participation in government

    J sustaining loyalty to deceased heroes

27 Based on the passage, Pericles probably believed that the Spartans —

    A ignored their civic duties

    B were thoughtless barbarians

    C lacked character

    D emphasized muscles over intellect

28 The capital of the Byzantine Empire was —

    F Constantinople

    G Rome

    H Jerusalem

    J Athens

29 Some historians attribute the final collapse of the Byzantine Empire in year 1461 primarily as a result of the empire's inability to —

A withstand the attacks by the Ottoman Turks

B maintain an efficient government bureaucracy

C remove divinely ordained emperors

D control costs of elaborate court ceremonies

30 If the Roman Empire had not collapsed, the feudal and manorial systems would have been less necessary in Old Europe. The statement is probably considered valid because —

F feudalism provided defense support for unprotected communities

G serfs began to demand more participation in governmental affairs

H noble vassals lost the right to hold privileged status and titles

J medieval churches became the seat of government authority

31 Two significant concepts related to Jewish religious and cultural identity are the —

A concept of reincarnation and Ramadan

B Torah and the Diaspora

C New Testament and the Four Noble Truths

D Koran and the code of bushido

32 Which statement about the social structure in Europe during the Middle Ages is most accurate?

F The nobles encouraged social mobility.

G The Catholic Church encouraged a classless society.

H Sharp class distinctions divided European society.

J Industrialization led to the growth of socialism throughout Europe.

33 While invasion by German barbarians represents a major factor in the fall of Rome, historians claim another equally important factor was due to —

A technological development

B imperial expansion

C economic decay

D spread of Christianity

34 Which economic system existed in Europe during the early Middle Ages?

F Free market

G Socialism

H Feudalism

J Communism

35 The Koran is associated with the practice of —

A Islam

B Judaism

C Shinto

D Buddhism

36 During the late 18th century, Thomas Malthus expressed alarm at the rapid growth of the world's population and doomed the future because the —

F food supply would exceed the world's population

G gold bullion would eventually cease to be valuable

H oceans would eventually evaporate leaving dry land

J world's population would exceed food supply

37 The Romans connected their empire by building a series of —

A canals

B roads

C railroads

D lighthouses

38 "Your words are wise, Arjuna, but your sorrow is for nothing. The truly wise mourn neither for the living nor for the dead. There never was a time when I did not exist, nor you, nor any of these kings. Nor is there any future in which we shall cease to be . . ."

The passage best reflects a belief in —

F ancestor worship

G the Eightfold Path

H reincarnation

J nirvana

39 Which of the following is the most significant factor explaining why Africa became a prime target for European imperialism following the termination of slavery in the late 1800s?

A Europeans discovered that training Africans for employment was more profitable.

B Technology enabled Europeans to extract and profit from Africa's resources.

C African nations demanded retributions for the devastation caused by slavery.

D Foreign governments wanted to atone for exploiting the African continent.

40 All of the following places exerted migration pull factors EXCEPT —

F river valleys during the Agricultural Revolution

G Mecca and Medina during the rise of Islam

H Constantinople after the fall of Rome

J Carthage after the Punic Wars

41 The Silk Road became a significant vehicle for China because it —

A provided access to trade with the Middle East and Europe

B represented the first paved road

C connected France and England

D created a quick desert route

42 According to the teachings of Confucius, the key to the successful organization of society is that —

F the rulers are democratically elected

G evil in humans is eliminated

H ancestor worship is discontinued

J individuals know and do what is expected of them

43 "It is better to do the work of your own caste poorly than to do the work of another caste well."

Of the following, which best summarizes the advice to peasants quoted above?

A Each person could choose his or her own occupation.

B Always strive to move higher up the ladder.

C Stick to your caste regardless of the consequences.

D Some castes do better work than other castes.

44 The tea ceremony, Kabuki theater, and haiku poetry remain important parts of contemporary Japanese culture. What does this suggest?

F Western culture influences contemporary Japanese life.

G The ideas of Confucius continue to dominate Japanese life.

H Social change remains a goal of Japanese society.

J Traditional customs and practices are valued in Japanese life.

45 Traditional art, music, and dance of Africa were most influenced by —

A religious beliefs and practices

B European Renaissance artists

C the effects of revolution

D South Asian artists

46 China remained isolated for many centuries because of its —

F natural harbors

G navigable river systems

H tribal conflicts

J mountain ranges

47 One similarity between the ancient African kingdoms of Egypt, Kush, and Ghana is that all of these kingdoms were located —

A in mountainous terrain

B in coastal areas

C on major trading routes

D in rain forest areas

48 Buddhism teaches that salvation is earned by —

F following the Ten Commandments

G worshipping Allah as the one true god

H commitment to spiritual self-discipline

J being baptized and confirmed

49 Which statement would most likely be spoken by a Hindu rather than a Muslim?

A "I am planning to go to Mecca next year to participate in the haji."

B "My belief in the teachings of the Koran are very important to me."

C "I always enjoy the feast that follows the end of Ramadan."

D "My good deeds today will serve me well in my next reincarnation."

50 A sociologist interested in ancient Indian culture would most likely focus on the study of —

F Hinduism and its influence on life in India

G India's development of nuclear weapons

H the parliamentary government of India

J the influence of agricultural production on India's gross national product

51 The two most significant cities of Islamic culture and religion were —

A Mecca and Medina

B Rome and Venice

C Athens and Sparta

D Sumer and Babylon

52 One reason the cultures of North Africa developed differently from the cultures of the rest of Africa was that these areas were separated by the —

F Congo River Basin

G Great Rift Valley

H Sahara Desert

J Arabian Sea

53 Compared to the Medieval period, the Renaissance shifted emphasis in all of the following areas except from the —

A emphasis on religion to secular affairs

B to portrait painting traditional biblical themes

C aristocratic birthrights to the self-made man

D Latin scriptures to classical Roman literature

54 On a map of the world, Asia is to Japan as Europe is to —

F Great Britain

G Austria

H the Netherlands

J Italy

55 In most societies, works of art and architecture generally serve to —

A satisfy the needs of the leaders

B limit the influence of religion

C reflect the values of society

D express opposition to the government in power

56 Which statement is an opinion?

F Russian athletes are successful because their nation has a superior culture.

G The gross domestic product (GDP) of Japan is greater than that of Thailand.

H China and India are the two most populous nations in the world.

J The majority of people in the Republic of South Africa is of African descent.

57 When land masses on a map are out of proportion with each other, it is known as —

A latitude

B longitude

C distortion

D incongruence

58 **The scale of a map determines the —**

F relation between a distance on the map and a distance in the real world

G key to the meaning of the various colors on the map

H method for determining the differences in the sizes of cities on the map

J time period of the boundaries that are represented on the map

59 **The latitude line exactly half-way between the North and South Poles is called the —**

A Prime Meridian

B Equator

C Tropic of Cancer

D Tropic of Capricorn

60 **Most traditional societies maintain social control and group cooperation through the use of —**

F subsistence farming

G regional elections

H democratic decision-making

J the extended family

61 **On a world map created during the European Middle Ages at the height of the Crusades, Jerusalem is located at the center of the map. On a world map created during the rule of the Third Reich during World War II, Germany is located at the center of the map. The reason for this difference is that —**

A during World War II, Jerusalem no longer existed

B maps often project the point of view of the map-makers

C during the Middle Ages, Jerusalem was the largest city, and during World War II, Germany was the largest country

D during the Middle Ages, Jerusalem was the center of the known world, and during World War II, Germany was the center of the known world

62 **Hieroglyphs are associated with which one of the following civilizations?**

F Ancient Egypt

G Mesopotamia

H Indus Valley

J Shang China

63 **All of the following are associated with ancient Greece EXCEPT —**

A Socrates

B Plato

C Julius Caesar

D Pythagoras

64 The end of the Punic Wars was a milestone for the Romans because they —

   F came under the control of Carthage

   G came under the control of Athens

   H became the superpower of the Mediterranean

   J had to share power with the Carthaginians

65 In western Europe during the Middle Ages, education declined as a direct result of —

   A the rediscovery of classical Greek civilization

   B the loss of the power of the Christian Church

   C the fall of the Roman Empire

   D the rise of the monarchy

66 A traditional society is most likely to —

   F discourage rapid population growth

   G reduce the influence of religious leaders

   H emphasize the well-being of the individual over that of society

   J accept social change on a limited basis

67 One result of the Neolithic Revolution was —

   A an increase in the number of nomadic tribes

   B a reliance on hunting and gathering for food

   C the establishment of villages and the rise of governments

   D a decrease in trade between cultural groups

68 Before West African civilizations had contact with Europeans, these civilizations developed —

   F art that included bronze, gold, and clay sculptures

   G economies that did not rely on trade

   H one system of government for the entire region

   J social systems that emphasized the nuclear family

69 Trade and wealth were prime motives for explorative contacts with Japan, but Japanese rulers during the 16th century were more favorably impressed with Westerners than other nations primarily because of the —

   A higher trade values

   B absence of slavery

   C change of monarchs

   D Jesuit missionaries

70 Longitude lines on a globe or a map stretch from —

   F east to west

   G the North Pole to the South Pole

   H the Equator to the Prime Meridian

   J Asia to the Americas

71 "From a little spark may burst a mighty flame."

—Dante

"Tall oaks from little acorns grow."

—David Everett

"The journey of a thousand miles begins with one step."
—Lao-tzu

Which conclusion is best supported by these quotations?

A All cultures are concerned with transportation and conservation.

B In all cultures, people make excuses for their mistakes.

C Geography is important to the development of all cultures.

D People from different cultures often view situations in similar ways.

Chapter 8

Answers and Explanations for Practice Test 1

Listed below are the answers to the Practice Test found in Chapter Seven.

1 B	11 A	21 B	31 B	41 A	51 A	61 B	71 D
2 H	12 F	22 G	32 H	42 J	52 H	62 F	
3 D	13 C	23 B	33 C	43 C	53 B	63 C	
4 F	14 H	24 G	34 H	44 J	54 F	64 H	
5 C	15 D	25 B	35 A	45 A	55 C	65 C	
6 F	16 H	26 H	36 J	46 J	56 F	66 J	
7 B	17 A	27 D	37 B	47 C	57 C	67 C	
8 J	18 J	28 F	38 H	48 H	58 F	68 F	
9 D	19 B	29 A	39 B	49 D	59 B	69 D	
10 H	20 J	30 F	40 J	50 F	60 J	70 G	

The field-test questions were numbers 4, 15, 20, 31, 38, 46, 55, 60, 67, and 71.

Learn From Your Mistakes

Review the explanations below for the questions that you missed, but don't just read the explanations! It's important that you also read the reason that the answer you picked is incorrect. If you understand why the wrong answers are wrong and why the right answer is right, you'll be less likely to make the same mistake when you take the real End-of-Course World History SOL exam.

What's more, try to figure out *why* you missed the questions that you missed. It's usually for one of two reasons: either you didn't understand the content of the question, or you were careless when reading the question and answer choices. If you didn't understand the content, review the relevant history and geography in the appropriate chapter in this book. If you were careless, slow down a little when taking the test and review the test strategies from chapters one and two of this book.

1   B **If you only know some of the words, eliminate the ones that you know do not describe people who follow herds of animals for survival.**

   A   Monotheists believe in one god; polytheists who believe in many gods.

   B   Nomads are people who don't stay in one place. Nomadic groups depend on the herds for survival.

   C   Ziggurats were pyramid-like structures built by the ancient Babylonians.

   D   Mesopotamians were people who lived in ancient Mesopotamia, an empire between the Tigris and Euphrates Rivers. Mesopotamians were generally not nomadic. They settled and built towns, cities, and farms along the banks of the rivers.

2   H **Merchants and artisans who fled to the cities during the Medieval period discovered that they had exchanged the security of the feudal manor for freedom from feudal taxes, rights of self-government, and expanded trade opportunities. Wars and conflicts were not eliminated by moving to the cities.**

   F   The flight to the city freed merchants and artisans from feudal labor compensated with excessive taxes, little or no pay, and total submission to the rich landowners. Escape from feudal taxes translated to more personal funds for basic subsistence. Unfortunately, urban independence also meant self-protection. In other words, they were on their own.

   G   Although self-government during the Medieval Age was a far cry from our current democratic process, at least merchants and artisans could develop laws and customs that facilitated their abilities to increase their standard of living.

   H   In no way did wars and conflicts end. In fact, vandals and barbarians were perhaps more rampant in the cities. Personal safety remained a key concern. Remember the moats and high turreted walls with massive gates surrounding King Arthur's castle?

   J   As merchants and artisans specialized and fine-tuned their skills, commercial opportunities expanded. Open markets and trade fairs opened avenues for production and trade.

3   D **If you're not sure which answer is correct, think logically and make the best guess. The question tells you that all three civilizations were built along rivers. You may know that these three locations are in different parts of the world, so you need to think of something that most likely would be true of three riverside societies with cultures that are probably different.**

   A   Monotheism didn't hit the Nile and the Tigris-Euphrates River valleys until the civilizations were already developed. As for the Huang Ho River civilizations in China, monotheism never really developed. Even if you don't know this, though, you should ask yourself the following question: Is there an answer that involves the fact that these societies developed alongside rivers?

   B   These are ancient civilizations, not Pittsburgh.

C Actually, none of these societies utilized anything resembling a democratic form of government. Instead, they were essentially monarchies run by clan leaders. Isn't there an answer that involves water?

D We know the three civilizations are located alongside rivers, so that takes care of the "natural waterways." Because of the rivers, we can presume there is fertile soil, and it doesn't seem unreasonable that there would be a mild climate as well. Done!

4 F **First observe the structure of the question and responses. Remember that Muhammad was the founder of Islam. This gives you the clue to the correct answer—the one that describes an association between a founder or creator, not necessarily a leader, and a concept or belief. This question has been simplified by relating all the responses to religion.**

F Similar to Muhammad, who was founder of Islam, Gautama Siddhartha founded Buddhism around 500 B.C. Distressed by the suffering and pain he observed, Siddhartha abandoned his noble trappings to wander as an ascetic. In a moment of revelation, he achieved enlightenment and became known as Buddha, or the Enlightened One.

G John the Baptist was not the creator or founder of Christianity. He was a disciple of Jesus who preached and promoted Christianity.

H St. Francis Xavier was a priest with allegiance to the pope as the leader of Catholicism. Although he was a prominent missionary who promoted education, his relationship to the pope is inconsistent with the stem. (St. Francis Xavier wasn't the *founder* of the pope.)

J The relationship of Moses to the Hebrews is also inconsistent with the stem because Moses was a leader who led the Hebrews to the Promised Land. Moses didn't found the Hebrews!

5 C **Read the quote carefully. If a noble knocks out another noble's tooth, he's going to lose his own tooth as well. But if a noble knocks out the tooth of a commoner, he pays a small fine. What can we conclude?**

A All men were not equal under this law. The noble's tooth is more valuable than the commoner's tooth.

B Fines were apparently preferable to corporal punishment when it came to hurting a commoner, but not when it came to hurting a noble. It would appear that the worse the deed (in the minds of the lawmakers), the worse the punishment.

C Yes! You just can't argue with this answer choice. There must have been a division between the social classes, or else the law would not have bothered to distinguish between commoners and nobles.

D This answer choice uses extreme language and is simply not true. First, we know that this particular part of the code provides that violence is punished with a fine if it is directed against a commoner. Second, we only have one portion of the code, and therefore, we can't make any generalizations about punishments for other types of violence.

6 F **If you only know about one of these civilizations, you can still answer this question correctly. If you don't remember anything about any of them except that they were ancient civilizations, you can eliminate at least a couple of answer choices.**

F The Phoenicians, the Sumerians, and the Maya all developed extensive writing systems. The Phoenicians laid the groundwork for the alphabet later used by the Greeks, the Sumerians used cuneiform, and the Mayas used hieroglyphs.

G None of these civilizations emphasized equality in education because none of these civilizations emphasized equality in anything.

H All of these civilizations were polytheistic.

J None of these civilizations encouraged democratic participation in government. Democracy was attempted for a while in ancient Greece and ancient Rome, and then didn't re-emerge until the European Enlightenment.

7 B **Use common sense. If you don't know the answer from your knowledge of India, think about the earliest civilizations in other parts of the world. If you can't remember anything about early civilizations anywhere on the globe, think through the answer choices logically.**

A India's earliest civilizations were not located in mountainous areas. It's cold there and farmland is virtually nonexistent.

B All known early civilizations were located in river valleys. River valleys provide water, vegetation, and transportation. Ancient Sumer, for example, was nurtured in the Tigris-Euphrates River valley. Ancient Egypt was built along the Nile River. Shang China grew along the banks of the Yellow (Huang Ho) River. So you shouldn't be surprised that early Indian civilization was located in a river valley as well. From approximately 2500 BC through 1500 BC, the early Indus Valley civilization thrived along the banks of the Indus River in what is now western India. Eventually, the civilization fell to Aryan tribes.

C Do deserts really help support early civilizations? The most important thing to sustain any society is water, and deserts don't have any of it.

D Steppes are large areas of grass-covered land that are typically very cold in the winter and very hot in the summer. Steppes are able to support nomadic groups, whose herds could graze on the fields, but not the development of major civilizations.

8 J **Answer choices F, G, and H all deal with environmental or agrarian issues. Only J doesn't have anything to do with fertility.**

F Placating (or appeasing) environmental gods was evidence of fertility in the religion of agrarian communities. Once communities learned to handle food surplus by storing food, seeding, and harvesting, farm chores could be planned according to seasons. Consequently, the deities (or gods) of the sun, moon, and seasons were very valuable. While not scientific, seasonal planning became more precise with agricultural advancements.

G The statement was evidence of the religion of an agrarian society. Drought or extreme weather easily doomed a community to starvation and loss of animals. In appreciation of the gods' favor, harvest was a time to pay sacrifices to the gods.

H Archeological finds of pottery and other artifacts bear pictures and symbols of nature, which were significant factors in crop fertility. So this answer choice *is* supporting evidence to the statement, therefore it's incorrect.

J The reference to successful hunts is a throwback to the earlier periods of hunter-gathering. Pictures memorializing hunts have been found on cave walls. This response, however, represents the wrong time and civilization. It doesn't support the statement, so it's the correct answer. (Remember to read the question very carefully!)

9 D The medieval church possessed the ability to expel or excommunication people, whereas modern states generally do not.

A Both medieval church and the state exercised the right to levy taxes. Priests and bishops did not receive salaries from the state, but they were supported by immense revenues from the landed estates owned by the church, not to mention the proceeds collected from the tithe, a tax imposed by the church on all land-owners. Today, the Catholic Church is among the world's largest real estate owners.

B Both the state and the church administered justice, although considerably different methods were used. Under state law, both modern and ancient, offend-ers were likely tried in civil courts ruled by kings or other officials. Religious offenders, including officials, were adjudicated under church or "canon" law. Citizens might also be tried in church courts for certain offenses such as sacri-lege, heresy, or theft of church property.

C Both the state and the medieval church hold the right to govern according to their particular brand of laws. The medieval church established laws and sys-tems to maintain order for the "community of all true Christians." Modern state laws govern the community of its citizens or residents.

D This is the best answer choice. While most modern states lack the authority to expel citizens, the medieval church could force expulsion or excommunication of members for violations of church tenets.

10 H Eliminate answer choices that you know are centered around other bod-ies of water.

F Shang China was centered around the Huang Ho River.

G The Mediterranean world is centered around the Mediterranean Sea.

H Mesopotamia literally means the "land between the rivers."

J Indus Valley civilization was centered around the Indus River in India.

11 A **The Agricultural Revolution was hugely important. If you don't remember anything about it, review your materials. Get rid of answer choices that don't make sense. You should be attracted to answer choices A and C, since they are direct opposites.**

A The Agricultural Revolution was so significant because it involved seed farming that led to food surpluses. Without food surpluses, virtually everyone in society has to grow his or her own food because there is none left over (a surplus) for anyone else. But with food surpluses, the farmers have extra food, which means that other people are freed to be productive in other kinds of ways. These other people built towns and cities and traded their own goods and services for the extra food of the farmers. Unlike nomads, farmers generally stayed in the same location year after year. This allowed the towns and cities around the farms to grow even larger.

B The Agricultural Revolution pulled people toward the river valleys since that's where the good farm land was located.

C As discussed above, the Agricultural Revolution is what made cities possible.

D The Agricultural Revolution did not eliminate political boundaries. If anything, it helped to create them. Since people increasingly began staying in the same place and building highly sophisticated cities and farms to support them, they became very territorial about the particular land on which they had made their investment. Boundaries between groups of people, therefore, became more important.

12 F **If you don't know the words in the answer choices, look them up. Ignore vocabulary at your own peril!**

F Ancient Athens is the only true example of a direct democracy. Currently, no direct democracies exist. In a direct democracy, each citizen votes on every issue, as opposed to representative or parliamentary democracy, in which citizens elect representatives to vote on every issue.

G Totalitarianism does not involve citizen participation in government. It involves only the totalitarian.

H A parliamentary democracy involves representatives who vote on behalf of those who elect them. This is quite different from a direct democracy, in which citizens vote on policy issues for themselves and which is the method of government described in the question.

J Absolutism does not involve citizen participation in government. It involves a single ruler who rules absolutely.

13 C **Try to remember everything that you can about ancient Greece. If you can't remember anything, you're in trouble because the test will include a lot of questions about Greece. Even if you remember only the basics about ancient Greece, you can still get this question correct because the wrong answer is very wrong!**

A It's true: The Greeks established a monetary system. The Greeks were the masters of commerce and trade, and a money system made doing business easier.

B The Greeks were big on trading and big on sailing.

C The Greeks *were* quite militarily advanced, especially the Spartans. Greece established colonies by attacking nearby territories.

D The Greeks had many gods, which means they were polytheistic.

14 H There were three types of people in ancient Greece: citizens, free people, and noncitizens. Try not to get them confused.

F Women who had no political rights but were not slaves were free, but they were not citizens.

G Citizens didn't include women, slave or not.

H Citizens were adult males. So even though ancient Greece was a democracy in which all citizens participated in government, to be a citizen in the first place, you had to be a member of the all-boys club.

J Citizenship in Athens was not the same as modern-day citizenship. It didn't include everybody, just nonslave males.

15 D Although the Romans imitated Greek culture in many ways, many of their concepts of justice and laws were original. Remember this item specifically addresses Romans, not Greeks, so this is your first clue.

A Ancient Roman cities were governed by a council and magistrates under the supervision of the governor. These levels organized the immense Roman Empire, which existed without any form of long-distance communication or transportation. Though more advanced today, many cities retain similar levels of local management through districts, councils, and magistrates.

B Yes, the Romans perfected the civil law for the administration of justice. One of the earliest examples of Roman law was the Twelve Tables, a very crude but functional system.

C Today, the Justinian Code continues to be studied in law schools throughout Europe as the basis of civil and administrative law worldwide.

D Thieves, murderers, and particularly bad debtors were harshly treated, possibly with hanging as punishment. Little thought was given to the rehabilitation of these criminals. Because of this, answer choice **D** is the right choice.

16 H Greek city-states were mini-kingdoms. Two Greek city-states dominated the region and battled each other for control. Get rid of any city in the answer choices that you know is NOT in Greece!

F Rome is not in Greece. It's in Italy. Roman civilization was impressive, but it was separate and distinct from Greek civilization.

G Carthage was in North Africa and was a mega-enemy of Rome.

H Sparta and Athens were the two dominant Greek city-states. Pericles, the Athenian ruler, pushed Athens into war with Sparta for total dominance of the region. The Peloponnesian War, as it was called, was catastrophic. Both Athens and Sparta were so weakened that they were conquered by the Macedonians.

J Carthage was in North Africa and therefore was not Greek.

17 A There are a couple of ways you can approach this question if you're not 100 percent sure of the answer. First, even if you're not sure who the leader of the Macedonians was, you can eliminate people who you remember led other empires. Second, focus on names of people that you remember were major world conquerors, even if you can't remember who they fought for.

A Alexander the Great spread *Hellinism* (a variation of Greek culture) throughout much of the known world. He conquered the Persian Empire and expanded his territory all the way into India.

B Julius Caesar was a major world leader, but was from Rome, not Greece.

C Hannibal was a Carthaginian, not a Greek. He tried to sack Rome but failed.

D Pericles was Greek and led it through its cultural golden age. But Pericles was not a military conqueror.

18 J Eliminate answer choices that describe something that was developed by someone other than Romans during the Roman Empire. Eliminate answer choices that don't describe something common in Western society. Eliminate answer choices that were developed well before or well after the rise and fall of the Roman Empire. Eliminate! Eliminate! Eliminate!

F The development of gunpowder was certainly significant, but the test writers probably wouldn't call it a "major contribution to society." Besides, gunpowder was developed after the fall of the Roman Empire in China. Romans didn't shoot each other in the back, they stabbed each other in the back!

G Socialism developed long after the fall of the Roman Empire.

H Monotheism developed long before the rise of the Roman Empire. And besides, the Romans were polytheistic.

J The Romans codified law. Their system (which they actually borrowed from civilizations in the Middle East) has served as a basis for much of modern-day Western law.

19 B Don't confuse ancient Rome with ancient Greece! Get rid of answer choices that apply to Greece or that remind you of Rome after the rise of Julius Caesar.

A Greece was a direct democracy. Rome was not.

B Rome was a republic. Its governing body was made up of two distinct groups, the Senate and the Assembly, much like the U.S. Congress. In a republic, the citizens have representatives who vote for them, rather than all of the citizens voting on every issue for themselves.

C *After* the rise of Julius Caesar, the Roman Republic dissolved and imperial Rome took its place. Julius Caesar became an emperor.

D *After* Julius, Augustus Caesar came to power and consolidated all the power in himself. Rome essentially became a totalitarian regime.

20 J   Learn about Confucius. Remember him. The test writers like him.

F   Confucius says: Put the group, especially the family, before the individual.

G   Confucius says: Respect authority.

H   Confucius didn't say much about reincarnation, but he was very concerned about how people behaved.

J   Confucius taught about conduct, specifically the responsibilities that individuals owe to others and to the groups to which they belong.

21 B   Even if you don't remember the term for the period of peace and prosperity in Rome, you should be able to eliminate some of the answer choices that clearly do not refer to Rome, or to peace and prosperity.

A   The golden age of Pericles was a time of prosperity in Athens.

B   The *Pax Romana* literally means "Roman Peace," a time when Rome became the capital of the Western world. Rome conquered huge swaths of land, but yet developed a relatively peaceful existence with its conquered peoples, allowing them to practice some of their traditional customs and allowing the arts to flourish. Ovid and Virgil were two of Rome's biggest contributors to literature, and they emerged during the *Pax Romana*.

C   The Edict of Milan was signed by Constantine after the fall of Rome and Emperor Nero's persecution of Christians. The edict was issued to end the persecution and led to the establishment of Christianity as the official religion of the Roman Empire.

D   The Reformation began when Martin Luther broke from the Roman Catholic Church. It started in Germany in the 1500s whereas Augustus' reign as emperor was around AD 30.

22 G   Only two answer choices—F and G—describe events that would logically lead to a decline of empire. So if you're guessing, you have a 50–50 shot.

F   If Rome suffered a series of military defeats in North Africa, this certainly would have contributed to its decline. The military defeats that were suffered, however, were due to invasions from the north.

G   The Roman government was corrupt and unstable. Leaders were assassinated and replaced; loyalties often were transferred or entirely absent. The lack of a clear plan and a hierarchy built on honor and trust contributed to the empire's downfall.

H   Slavery was not abolished in the empire, but even if it were, it likely wouldn't have contributed to the empire's decline.

J   Continued acceptance of traditional religions had no great impact on the decline of the Roman Empire. The Roman government's policy was to allow local regions to practice local religions, which contributed to stability.

23 B If you don't know the name of the eastern portion of the Roman Empire, at least try to eliminate empires that you know weren't the eastern portion.

A The Persian Empire preceded Christianity entirely.

B The Byzantine Empire was the eastern portion of the Roman Empire, centered in Constantinople.

C The Mongol Empire was established in China and eventually in parts of Persia in the thirteenth and fourteenth centuries.

D The Gupta Empire was all the rage in India in the fourth through the sixth centuries.

24 G Either you know this one or you don't. A few questions on the test will just be straightforward questions of knowledge.

F Londinium was settled by the Romans, but Britain was not known as Gaul.

G France was known as Gaul.

H Germany was not settled by the Romans but rather was the source of some of the invading tribes that eventually brought the downfall of the empire.

J Russia was not settled by Romans.

25 B Remember: Primary sources are sources that provide first-hand accounts of an experience. Get rid of any answer choice that refers to a person writing about a topic that he or she was not involved in personally.

A A twentieth-century historian could not have been personally involved in the Peloponnesian War.

B A citizen of Carthage who personally experienced the Punic Wars would certainly write a primary source about the nature of the Punic Wars.

C A television account of the Roman Empire would involve lots and lots of sources, but the account itself would be a secondary source.

D Correspondence between two medieval monks might be a primary source of information about life in a monastery during the Middle Ages, but their correspondence would not be a primary source about the philosophy of Aristotle, since they would have lived centuries after Aristotle.

26 H According to the summary of the speech, Pericles seems to believe that the essence of a city's survival is to ensure citizen participation in government.

F As a person committed to the government of Athens, Pericles does not necessarily ignore strong military defense, but more important in this statement is his emphasis on the role of citizenship in a democratic government. Spartans would probably share this interest only if military obligations were emphasized.

G Pericles's oration advocates precisely the opposite of this answer choice, clearly qualifying citizen distinction as a reward of "public service" rather than a reward of privilege, birthright, or upper class status. He clearly defines distinguished citizens as those in public service.

H Pericles applauds democracy, where everyone has a role. According to Pericles, democracy is a system that utilizes the talents and minds of many, without "loss of manliness." Citizenship, rather than bravery, is the key to manliness, according to Pericles. This is the best—and the correct—answer choice.

J In no way is Pericles suggesting that deceased heroes not be honored. He probably qualifies honors exclusively for deceased heroes considered prominent public servants. In any case, sustaining loyalty to deceased heroes, according to the passage, is not the essence of a city's survival.

27 D **The statement seems to say that the Spartans "from early youth are undergoing laborious exercises to make them brave," emphasizing muscles over intellect. Answer choice D makes the most sense.**

A The passage does not imply that Spartans ignored their civic duties. More important is the emphasis given to military capability, perhaps as the core of civic duty.

B This response is incorrect because the statement does not imply that Spartans are barbarians. Such a conclusion might be inferred, but it is not necessarily a valid conclusion.

C The matter of character is not discussed in this statement. From another perspective, character, in the Spartan sense, might well be determined according to physical attributes.

D Pericles's statement implied that Spartans perceived physical fitness rather than the proclivity for beauty as a validation of manliness. Pericles reminded the audience that Athens's appreciation for beauty also represents manliness in the performance of civic duties.

28 F **You need to know the major cities that are integral to world history. The capital of the Byzantine Empire is one of these cities. If you missed this question, review the material regarding the rise of the Byzantine Empire.**

F Constantinople became the capital of the Byzantine Empire under the leadership of Emperor Constantine. Built as a "second Rome," Constantinople became the center of Christianity as well as trade and commerce after the fall of Rome.

G Rome was, of course, the capital of the Roman Empire, not the Byzantine Empire.

H Jerusalem was an important city to the Byzantine Empire because of its central place in Christianity, but Jerusalem was controlled by Muslims during much of the Byzantine Empire's existence.

J Athens was the center of culture in ancient Greece, but had lost its luster by the time of the Byzantine Empire.

29 A The Byzantine Empire could not survive the unrelenting attacks by the Ottoman Turks. Following two months of attacks and bombardment in 1461, the Byzantine Empire succumbed to the Turks' control. The Turks' invasion was one of many that threatened the empire.

A Long before the final defeat by the Turks, the divided Byzantine Empire endured one conflict after another from the west. By 1461, the Turks had conquered all the outlying pockets of the Byzantine authority and finally phased the weak empire out of existence. This is the correct response.

B This choice is almost the opposite of the right answer. The Byzantine Empire developed one of history's most extensive and elaborate bureaucracies. The government was divided into departments such as the chancellor in charge of all state correspondence, just like our postal system. Military governors ran the provinces, known as themes. Each department maintained specific functions, but the centralization of power was in the hands of the emperor.

C This is probably a good response applicable to most ancient governments, but not so for the Byzantine Empire. The emperor was the head of the church as well as of the secular state, by *divine ordination*, quite similar to divine right. However, by the same power of appointment, the emperor could also be removed. The justification was that if the emperor stirred the populace to rebellion, this was a divine sign to change emperors. Of course, though, an unsuccessful rebellion against an emperor was considered treason against God as well as the emperor.

D The Byzantines were famous for elaborate ceremonies and majestic rituals. They were probably extravagant, but not sufficient to bankrupt a thriving empire during its golden years. Financial problems, however, were devastating as the Byzantine Empire declined. The Byzantines focused on education, charities, hospitals, and welfare for the poor. The poor were basically a professional class represented by a company structure comprised of beggars who controlled the best stations in the city.

30 F It helps to know something about feudalism for this problem. In short, feudalism provided support for unprotected communities, making answer choice F the right one.

F If you checked **F**, you are correct because the disintegration of Roman power left western Europe without any capable authority for maintaining and protecting trade and valuable land. Communities were left with inadequate defense and economic support. Consequently, they looked to military chieftains and large landowners in their vicinity for protection. The landowners asked in return for obedience, services, and the payment of rents. So began the feudal and manorial systems.

G This answer choice is incorrect because serfs or peasants demanded no rights of independence or government participation during the Middle Ages. Their appeal to landowners was for food and safety, which they paid for with their sweat and obedience.

H Because power was concentrated among the monarchy, the only right lost by nobles was that of waging private wars. They remained subjects of the monarchy with authority to manage the king's estate and lived off the fat of the land. With their noble title and status, they comprised the class known as privileged aristocracy. Today, Europe is swamped with dukes, counts, or barons, all titled aristocrats. They are not necessarily wealthy, but they are titled because they are mostly descendants of privileged and powerful landowning aristocrats.

J This is an incorrect answer, and it's a bit offbeat. Politics and economics definitely affected the authority of the medieval church, but the downfall of Rome did not cause the church to supplant the government. Its role was altered, and although the church gained power in some areas, it did not produce the feudal system.

31 B **If you don't know the items in the answer choices, it's time for you to do a little reviewing of the major religions. We want an answer choice that tells us about Jewish faith and culture.**

A Ramadan is a Muslim holy month. Reincarnation is a Hindu belief.

B The Torah is the first five books of the Old Testament and is central to the Jewish faith. The Diaspora (meaning *completely scattered*) was the scattering of the Jews in the first century AD throughout Europe and the Middle East. Even if you know only one of these terms, this is the answer choice to pick.

C The New Testament is central to Christians and the Four Noble Truths are essential to Buddhists.

D The Koran is the central book of Islam. The code of bushido is the guide of conduct for Japanese Samurai warriors.

32 H **If you remember that feudalism existed in Europe during the Middle Ages, it would help. Alternatively, if you remember that during the Middle Ages, Europe was a traditional society, you'd get the right answer for the same reason, as long as you remember the general characteristics of traditional societies.**

F Nobles did not encourage social mobility. The nobles had no reason to encourage social mobility because they benefited if the serfs remained loyal to them.

G The practices of the Catholic Church did not lead to the development of a classless society. The Catholic Church is hierarchical and supported a class-based society.

H European feudalism was marked by sharp class distinctions. Look up *feudalism* in this book for a detailed description if you missed this question.

J This answer choice is way out of the time period. Industrialism occurred well after the European Middle Ages.

33 C **The only answer choice that makes any sense in the context of the question is C, economic decay. Would any of the other choices lead to the downfall of an empire?**

A Answer choice **A** is incorrect because it was the *lack of technological development* rather than technological development that contributed to Rome's downfall. Historians highlight the fact that Rome was technologically retarded. The incentive for developing new technology was offset in part because of slavery and the free-flowing lifestyle.

B Imperial expansion was not a direct cause of Rome's decline. In many ways, imperial expansion masked the basic weaknesses in Rome's structure for a long time.

C Economic decay was definitely the other demon that contributed to the downfall of Rome. Low productivity, inflation, and worthless coinage all substantially weakened Rome's economy. With a shrinking economy, facilities deteriorated and many industries were forced to shift from the cities to villages. The decline of trade and shrinkage of state revenues signaled the end of a golden Roman era.

D The spread of Christianity did not cause the fall of Rome. While Christianity did not gain universal acceptance in Rome, several rulers were converted and halted the persecution of Christians.

34 H **If you didn't get this question correct, review major economic systems as well as the philosophies behind them. If you don't know the terms in the answer choices and the approximate time periods in which they were developed, it'll come back to haunt you again and again. So make sure you review them now in this book and in your materials!**

F Free-market capitalism didn't develop until after the Middle Ages.

G Socialism didn't gain a lot of steam until the nineteenth and twentieth centuries.

H Feudalism is so tied to the Middle Ages that you should think about it as soon as you hear the words "Middle Ages." Look up *feudalism* in this book if you don't know what it was.

J Communist economies were developed in response to the exploitation occurring in nineteenth-century capitalist economies.

35 A **You only need to know about one of the items on the list in order to answer this question correctly, and you definitely should know about the Koran (also spelled Qur'an) even if you don't know about the other three items.**

A The Koran is the sacred text of Islam composed of writings accepted by Muslims as the revelations Allah made to Mohammed through the Angel Gabriel.

B The Koran has nothing to do with Judaism.

C If you picked this answer, you really need to consider spending some more time reviewing major world religions.

D Nope. Buddhism is entirely unrelated to the Koran. Study major world religions!

205

36 J **Think about one of the biggest issues today, one that people are still worrying about. It has nothing to do with having too much food, or having the oceans dry up. Malthus predicted that human population would exceed the planet's food supply, answer choice J.**

F Excessive food supply is the opposite of Malthus's prediction. The key word in the stem is "doomed"; therefore, the response illustrates an adverse outcome. With modern technology and the world's starvation problems, excessive food would rarely be considered a negative or adverse result.

G Gold bullion devaluated? It's possible, even likely, but it's unrelated to Malthus's theory.

H This is an incorrect response because it does not address Malthus's argument.

J This is a good answer. In his book, *An Essay on the Principle of Population,* Malthus warned the leaders that increasing birth rates would lead to an insufficient food supply to feed the world. His predictions continue today as the Malthus theorem among demographers. Malthus's predictions receive greater weight today because of the impact of environmental hazards on agriculture and the increasing births in underdeveloped countries. A major debate among economists, demographers, and health professionals is how to balance food and supplies for the world's six billion (and growing) inhabitants.

37 B **You should be able to eliminate answer choices simply by understanding the time periods involved. Pick the answer choice that logically would "connect" an empire.**

A While the Romans built aqueducts and canals, these were not the primary method by which places in the empire were connected.

B The Romans were famous for their roads. Their roads were so sophisticated that we still use the same fundamental technique of road-building. These roads allowed the empire to be controlled from a centralized location.

C Railroads weren't invented until the Industrial Revolution, more than 1,000 years after the fall of Rome.

D While the Romans were great shipbuilders, lighthouses did not tie the Roman world together.

38 H **Make sure you read the passage carefully, looking for key words that are part of the definition of one of the beliefs in the answer choices.**

F This passage is not about ancestor worship because the passage explicitly states that "the wise mourn neither for the living nor for the dead." If you were worshipping your ancestors, you'd be thinking about dead people all the time.

G Following the Eightfold Path is a practice of Buddhism. The Eightfold Path lists the basic rules of conduct and thought. But this passage isn't about conduct, it's about the nature of existence.

H This is the best answer choice. If "there was never a time when I did not exist, nor you, nor any of these kings," then that means everyone always exists. But yet we know from experience that individuals die. We know that people die, but people don't die. What is the way out of this paradox? Reincarnation. An individual life is just a link in the chain of an everlasting soul that is reborn again and again—one soul with lots of separate identities. Reincarnation is a central belief of Hinduism that makes the caste system possible.

J Nirvana is the state of complete bliss that one hopes to achieve by following the Four Noble Truths and the Eightfold Path in the Buddhist religion. This passage is not about the state of complete happiness and peace. It's about reincarnation. And according to the Hindu religion, sometimes people are reincarnated into very unhappy lives, depending on how they lived their previous lives.

39 B If you're guessing, eliminate answer choices that seem unlikely. Choose the answer choice that makes the most sense.

A Europeans had absolutely no interest in Africans' welfare. European nations focused on mercantilism.

B The Industrial Revolution introduced new technology capable of creating more profits with less overhead. Anti-colonial attitudes toward Africa vanished as nations discovered that new technology could be utilized to extract and transport Africa's minerals and raw materials.

C Retributions, amends, or apologies were rarely part of negotiating or addressing any evils perpetrated against Africa though slavery. Frequently, nations would cloak imperialism under idealistic perceptions of Africans as the "white man's burden."

D Remember that the emphasis during the 1800s was on wealth rather than atonement.

40 J Pull factors are things that make you want to migrate to a particular place. These are different from push factors, which make you want to move away from a particular place. Three of the answer choices involve places that people would want to migrate to during the time mentioned in the answer choice. The right answer is the place that a person wouldn't want to migrate to during the time mentioned in the answer choice. These types of geography questions rely on a knowledge of history, which is why the geography questions and history questions are intertwined on this exam!

F River valleys were attractive places to move to during the Agricultural Revolution. Not only was the farm land ideal in the river valleys, but because of the surplus of food, towns developed along the rivers as well. The river valleys became the centers of civilization after the Agricultural Revolution!

G Mecca and Medina are holy cities in the Islamic faith. After the rise of Islam, they became incredibly important cities for faithful Muslims to visit.

H After the fall of Rome, European culture shifted to Constantinople. It became the center of an empire, the center of trade and the arts, and the center of Christianity. Definitely a place where people migrated!

J Thanks to the Romans, Carthage was in ruins after the Punic Wars. Its importance on the world stage plummeted.

41 A **The Silk Road connected Asia to the Middle East and Europe. It got its name because silk was a hot commodity in the west, and the nation that made the most silk in the east was China.**

A China, a relatively isolated country, traded some of its precious commodities, especially silk, along the Silk Road, an extremely long road extending out of central China to the Middle East, where it linked with other roads to Europe.

B Paved roads did not appear until the nineteenth century.

C The Silk Road came nowhere near England or France. Even if you don't know where the Silk Road was located, you can eliminate this answer choice by thinking to yourself: *Why would a passageway between England and France be called the Silk Road?*

D The Silk Road went through some desert territories. However, it was not fast. The journey took several months.

42 J **If you don't know Confucius, look him up in this book. You mustn't be confused about Confucius.**

F Confucius has been dead for more than 2,000 years, long before democracy hit Asia.

G Confucius wasn't concerned about eliminating evil; he was interested in instilling responsibility and respect. Even if he was concerned with evil, this answer choice is stated too strongly. Beware of phrases such as "must be eliminated." Test writers are more likely to use less extreme phrases, such as "should be weakened."

H Confucius believed that people should respect their ancestors, both living and dead. **Remember:** People have been admiring Confucius for more than 2,000 years. Wouldn't it be ironic if he didn't think he should be respected by later generations?

J Knowing and doing what is expected of you is the essence of Confucianism. Since you will be expected to know Confucius for the World History test, you will be *practicing* what he taught if you *learn* what he taught.

43 C **This person doesn't want anybody to be doing anything outside his or her caste. In other words, the person could have said, "Stay in your own caste, no matter what the consequences."**

A Each person has to stay in his or her own caste.

B This answer choice has nothing to do with the quote.

C According to the quote, you should stay in your caste, no matter what.

D This quote doesn't tell us anything about the upper castes or happiness. All we know is that the speaker believes it is better to work poorly in your caste than to work well outside it.

44 J You don't have to know about the tea ceremony, Kabuki theater, or haiku poetry to answer this question correctly. For this particular question, you only need to know how to read carefully in order to gain a point. Look closely at the words in the question. The test writers tell you that these activities "remain important parts of Japanese culture." If they remain important, then these customs, whatever they are, were important in the past. And if you understand that, then all you have to do is pick the answer that you just can't argue with!

   F All of these cultural practices originated in Japan, not in the West. Even if they did originate in the West, that still wouldn't explain why they *remain* important in Japanese culture.

   G Confucius taught proper conduct, not cultural practices involving theater and poetry. What's more, the words used in this answer choice seem too extreme. Even if you don't know how these practices originated, you should realize that they don't *dominate* Japanese life. Japanese life is dominated by work, education, and obligations to family. It is *influenced* by cultural practices such as the tea ceremony, Kabuki theater, and haiku poetry.

   H If change is a goal, then why would anything *remain* important? If social change is important, then it is likely that social customs would change accordingly.

   J Yes! This answer choice is almost impossible to argue with, given the wording in the question. These three social practices *remain important* because the Japanese *continue* to value *traditional* customs and practices. Or, in other words, if the Japanese didn't continue to value traditional customs and practices, then they wouldn't have used the words "remain important" in the question.

45 A The key words here are **Africa** and **most influenced by.**

   A **Remember:** The role of religion in many societies has been extremely significant. Many traditions, rituals, and aspects of daily life have been greatly influenced by religion. In the case of African religions, art, music, and dance play important roles in rituals and ceremonies.

   B Why would traditional music and art in AFRICA be most influenced by what's going on in Europe? If so, it wouldn't be traditional African music and art, it would be borrowed.

   C Revolutions change things. *Traditional* art and music are the ones that survive through changes.

   D This is Africa we're talking about, not South Asia!

46 J First, think of the geographical features of China. Then focus on the words "remain isolated for many centuries." Which geographical features would help to keep a country isolated? Even if you can't remember anything about China's geography, pick an answer choice that makes sense!

   F If a country has a lot of natural harbors, that means it is an ideal place for a strong shipping industry. If a country has a strong shipping industry, it wouldn't remain isolated since the ships would be going to and coming from other places! This answer choice doesn't make any sense.

G If a country has navigable rivers, it would be easy to take a boat up the rivers and into the interior of the country. But if it's easy to get into the interior of China, then China would be less likely, not more likely, to remain isolated. So this answer choice doesn't make any sense.

H To be sure, tribal conflicts would have a negative impact on transportation and communication. However, China does not have tribal conflicts. In fact, it has been a united country for thousands of years. You should also know that China is the most populous nation on earth. If tribal conflicts there were severe, it's unlikely that such a huge civilization would have developed!

J This is the best answer. Mountain ranges are natural barriers to transportation and communication and keep the places on both sides of the range relatively isolated from each other. In China, mountain ranges to the north and the west kept it separated from the rest of Asia, so China developed relatively independently. The Silk Road in the east allowed some trade with other nations, but it wasn't until sea travel became advanced enough to permit regular international travel that China was opened up to the rest of the world.

47 C **Even if you know only two of these ancient kingdoms, you can still get this right. Start eliminating answer choices that don't apply to the kingdoms you know about.**

A None of the kingdoms was in the mountains. Even if you only know about one of them, strike this answer choice.

B Only the ancient Egyptian kingdom was on the coast.

C This is it! The Egyptian kingdom was on the Nile and the Mediterranean. The other three were on major land routes going from North and West Africa to East and southern Africa. It also makes sense that a kingdom would grow strong because it traded with other kingdoms (and taxed them when it could).

D None of the four kingdoms was in a rain forest area.

48 H **If you know about Buddhism, which you should, then you can simply pick the correct answer choice. If you don't, then you should eliminate answer choices that relate to religions that you do know about.**

F Salvation earned by following the Ten Commandments is central to Judaism. The Ten Commandments are also important to Christians, but they do not believe that salvation is earned through practicing them.

G Worshipping Allah as the one true god is central to Islam.

H Buddhists believe that complete peace can be reached if one frees oneself from the distractions of the physical world and focuses on spiritual self-discipline. Salvation is not earned through any means other than commitment to spiritual self-discipline.

J Baptism and confirmation are central to the Christian faith. According to Christian doctrines, however, baptism and confirmation are not the sole ways of earning salvation.

49 D You can answer this question correctly if you know nothing about Hinduism, and a lot about Islam. You can also answer this question correctly if you know nothing about Islam, but something about Hinduism. But you will only be able to guess correctly if you know nothing about either.

A Mecca is the most important city of Islam, not Hinduism.

B The Koran is the most important book of Islam, not Hinduism.

C Ramadan is an important holy month in the practice of Islam, not of Hinduism.

D Reincarnation is a central belief of Hinduism, not Islam.

50 F The clue phrase in this question is "sociologist interested in Indian culture." You need to ask yourself the following question: Which answer choice tells us something about the culture of India?

F Hinduism is the dominant religion in India. The traditions and beliefs of a certain religion often tell us a lot about the culture in which that religion is practiced. Even if you don't know that Hinduism is the dominant religion in India, you should still choose this answer because it refers to the "influence on life in India."

G A sociologist is going to be interested in practices that affect daily life in India, not in military matters or high technology, except to the extent that they influence daily life.

H A political scientist would be interested in the type of government in India, but a sociologist would rather know about things that directly influence daily life.

J An economist or investor might be interested in the influence of agricultural production on India's GNP, but a sociologist wouldn't be. A sociologist would want to know about the influence of agricultural production on daily life and culture.

51 A Mecca and Medina were the two most significant cities of Islamic culture and religion. If you didn't know that, you could probably use POE to help eliminate some other answer choices.

A Medina and Mecca were important centers of Islamic culture and religion. The structure of the Islamic community during the early days of the religion was comprised of many different nomadic tribal units that maintained their identities within small towns along caravan routes. Mecca represents the foundation of Islam. To make at least one pilgrimage in a lifetime to the shrines of Mecca is a sign of devotion to Islam. In addition to the Prophet Muhammad settling in Medina shortly before his death, Medina achieved fame as a trade center.

B Rome and Venice are incorrect because neither had an association with Islam.

C Athens and Sparta were important mainland cities among the Greek city-states.

D Sumer and Babylon were associated with the early Mesopotamian civilization, not Islamic culture.

52 H Your review for this test most definitely should include an atlas. Go find one if you missed this question. Open it up to a topographical map of Africa. Now, while looking at it, think about this: What separates North Africa from the rest of Africa?

F The Congo River Basin is in Central Africa. It does not separate North Africa from the rest of Africa.

G The Great Rift Valley runs down the spine of central Africa, separating the great forests of the Congo Basin and Zambezi River valley from the East African nations of Mozambique, Tanzania, Kenya, and Uganda. It's nowhere near North Africa.

H This is the best answer. The Sahara Desert dominates the landscape of North Africa, separating it geographically and culturally from the rest of Africa. The difference is so dramatic that the rest of Africa is simply referred to as *sub-Saharan Africa*. Although several West African kingdoms established regular Saharan trade routes through which they amassed tremendous wealth, for the most part, the desert culturally separated North Africa from the rest of Africa. As a result, North African kingdoms developed cultural links with the Middle East, while the rest of Africa developed more independently.

J So you say you don't even see the Arabian Sea on your map of Africa? Don't blame the mapmakers—they did their job. The Arabian Sea is way off to the east of Africa, between Saudi Arabia and India.

53 B The most important clue in this question is "Renaissance," which means a rebirth. Therefore, the correct response must address characteristics indicating a rebirth or newness.

A A is an incorrect response because the Renaissance indeed shifted emphasis from religion to the secular world of current affairs and business. In fact, even more attention was given to the quality and design of houses, clothes, and definitely pleasure. Secularism rejected the false spirituality and asceticism prevalent during the medieval time.

B This is the correct answer because the shift was the reverse—the Renaissance shifted artistic themes from the Medieval emphasis on biblical themes to the Renaissance feature of human emotions, and body form, adorned with color background scenes.

C The Renaissance certainly reversed the old feudal fixation on wealth or aristocratic birth as the key to success. Therefore, this isn't the answer choice we're looking for. Renaissance social class was defined by social and economic criteria, rather than the high birth criteria of the Medieval Ages. The emphasis on scholarship and commerce enabled the average person to become successful as a "self-made" person.

D The Renaissance began to emphasize classical literature rather than biblical scriptures of the medieval time. Therefore, **D** is also incorrect. The revival of Roman classical literature stimulated searches for classical literary texts in monasteries and other sites.

54 F **All four nations in the answer choices are part of Europe. Therefore, you need to look for the answer choice that is unique from the other three in a way that makes it similar to Japan and its relationship to Asia.**

F Japan is an island nation located off the coast of mainland Asia; Great Britain is an island nation located off the coast of mainland Europe. Bingo!

G The Netherlands, while part of Europe, is not an island nation.

H Austria is not an island nation. In fact, it's landlocked!

J On a map, Italy looks like it's kicking an island (Sicily), but it is not itself an island.

55 C **Remember: You should always look for the answer choice that you just can't argue with. Often, this answer choice is the one that uses generalized, "soft" words. In general, what can you tell about a society by looking at its art and architecture?**

A Sometimes leaders commission buildings and pieces of art, but usually the government is not involved in the lives of all or even most of the artists and architects (except in extreme cases, such as fascist or dictatorial states).

B In most religions, art and architecture have significant roles. Think of all of the elaborate and ornate churches, temples, and mosques in the world and of all of the paintings and sculptures that have religious significance.

C Taken as a whole, the art and architecture of a society give us an idea of what that society is or was like. Paintings, for example, often give us scenes from everyday life or reflect the hopes of what life might one day be like. As for architecture, it gives us an indication of what types of buildings and what rooms within those buildings are central and important, so it gives us an indication of the society's values and lifestyle. This is the most general answer choice, and it is almost impossible to argue with!

D In many societies, artists use their talents to express opposition like few others can. But this is certainly not true of *most* art in *most* societies.

56 F **An opinion is something that can be argued, but not proven. A fact, on the other hand, can be proven.**

F This is an opinion. In fact, this statement involves at least a couple of opinions and a lot of ambiguity. *Success* is ambiguous because people disagree on what success is. Even if there was agreement, however, it is still an opinion. No yardstick or calculator can tell us for sure which human event is the cause of another human event.

G This is a fact. It can be proven. All we have to do is measure the GNP of the two nations to determine which one is bigger. The people who thought Thailand's GNP was bigger would then be *proven* wrong.

H This is a fact. It can be proven. We can count the number of people in every single country on Earth. It might take a while, but it can be done. Then all we have to do is compare the population totals to determine which two nations have the greatest number of people. It turns out that it is China and India.

J   This is a fact. We can count the number of African-descended people in a country. We can count the number of non–African-descended people in a country. We can compare the two numbers and determine which one is larger. Therefore, comparing the populations is certainly *not* an opinion.

57 C **If you missed this question, review the characteristics of maps and map-making in the Geography Skills section.**

A   Lines of latitude are those that extend east and west across a globe or map.

B   Lines of longitude are those that extend north and south across a globe or map.

C   Distortion is the term that map-makers use when a map projection makes a territory appear to be larger or smaller in comparison with other territories than it actually is.

D   Two things that are incongruent don't go well together. It is not a map-maker's term, but rather a term included to distract you.

58 F **All good maps have scales. If you missed this question, review the characteristics of maps and map-making in the Geography Skills section.**

F   The scale of a map tells you how to compare distances on a map with distances in the real world. For example, if you look at a map of Virginia, the scale may tell you that 1 inch equals 25 miles in the real world. That means that towns that are 4 inches apart on the map are about 100 miles apart on the ground.

G   If a map uses a bunch of colors to distinguish between territories, a *key* or a *legend* would indicate what the colors mean, but this key or legend would not be the scale.

H   Sometimes maps use different shapes to distinguish between towns of different sizes. For example, on a map of Virginia, Richmond might be represented by a large dot with a circle around it, while Charlottesville might be represented by a small dot. A *key* or a *legend* would help you determine what the dots represent, but this key or legend would not be the scale.

J   It's important to look to see if a map has a date since boundaries change with time. This is especially true if you are looking at a historical map. But this date is not the scale. It is simply the date!

59 B **Review the information in the Geography Skills section if you missed this question.**

A   The Prime Meridian runs north and south and connects the North Pole to the South Pole. Therefore, it can't be halfway between them.

B   The Equator is exactly halfway between the North Pole and the South Pole, running east and west at the center of the globe. Climates along the equator are tropical.

C   See explanation for choice **D**.

D   Both the Tropic of Cancer and the Tropic of Capricorn are lines of latitude like the equator, but they are not exactly halfway between the North Pole and the South Pole.

60 J We hope you didn't miss this question because you don't know what a traditional society is. The term "traditional society" is an important term that you need to know. Only two answer choices describe characteristics of a traditional society, and only one answer choice describes how "most traditional societies maintain social control and group cooperation."

F Subsistence farming is characteristic of most traditional societies. However, subsistence farming isn't a method by which most traditional societies maintain social control and group cooperation. It's merely the label social scientists have given to farming methods that don't result in an agricultural surplus.

G Regional elections are not common in traditional societies. Even if they were common, it's highly unlikely that regional elections would be a significant method by which the society would maintain social control and group cooperation. Where elections exist, they usually are methods by which people choose their representatives. Typically, elections occur only once or twice a year, if that. Therefore, given their limited use and occurrence, elections—regional or otherwise—are unlikely to be methods by which societies maintain social control and group cooperation.

H Democratic decision-making rarely occurs in traditional societies. This answer choice is also very similar to answer choice **G**, since democratic decision-making usually occurs by way of an election. To be sure, it makes sense that democratic decision-making might be a method by which societies might maintain group cooperation, but in the case of traditional societies, most decision-making is class-based and rank-based. In other words, the leaders of a group make decisions for the rest of the group; the leaders of the family (parents) make decisions for the rest of the family.

J This is the best answer choice. Most traditional societies maintain social control and group cooperation through the use of the extended family. Generations within the same family often live and work together. Social control and group cooperation are strengthened by this system because an individual's identity is tied to the identity of his or her family, and as a consequence, individuals rarely stray from doing what is beneficial to and expected of the entire family. Typically, when individuals leave their extended families in large numbers, traditional societies begin to unravel.

61 B Maps are created for a variety of reasons. Look at the information in this question carefully. Even if you're not sure why Jerusalem and Germany would be located at the center of these maps, think about the reasons why mapmakers would choose to put certain parts of the world in the centers of their maps!

A This doesn't make sense. Jerusalem existed during World War II and still exists.

B This answer choice makes sense. The point of view of the mapmaker (or the person who is paying the mapmaker!) is extremely relevant to the look and feel of the map. During the Middle Ages, Christianity dominated Europe. Christian doctrines and history affected everything. So Jerusalem, perhaps the most important city in Christianity, was put at the center of the map to represent its central importance to the religion. During World War II, Germany considered itself to be the center of the civilized world, and so some German mapmakers put Germany at the very heart of the map. Of course, the Earth is a sphere, and so its surface doesn't have a center. Nevertheless, when projecting the sphere onto a flat surface, mapmakers must choose which part of the sphere goes in the middle and which parts go in the corners.

C Mapmakers usually do not put the largest city or country at the center of the map unless they are doing so for a very particular purpose. Second, Jerusalem was not the largest city during the Middle Ages, nor was Germany the largest country during World War II.

D To say a place is at the center of the known world doesn't make sense, because our world is a sphere.

62 F **This one is hard to guess at if you have absolutely no idea what hieroglyphs are. Even if you remember that they are an ancient writing system, all the civilizations in the answer choices are ancient! If you know enough about hieroglyphs writing to visualize it, try to do so. Do you remember anything that reminds you of one of the civilizations in the answer choices?**

F Hieroglyphs were an advanced writing system used in ancient Egypt. The system used a series of pictures to represent words and ideas. Hieroglyphic inscriptions can be found on the tombs of the pharaohs and other significant pieces of Egyptian architecture.

G Sumerians, who were the first to settle in Mesopotamia, had a writing system known as *cuneiform*.

H The Indus valley civilizations were semi-nomadic and not big on writing.

J Shang China developed an extensive writing system that allowed many people to become literate. But it wasn't called hieroglyphs!

63 C **Three of the people are associated with ancient Greece. Can you spot the Roman?**

A Socrates was a Greek philosopher.

B Plato, too, was a Greek philosopher.

C Julius Caesar was a Roman. Try not to confuse the Greeks with the Romans.

D Pythagoras was a Greek mathematician.

64 H **If you don't remember the details of the Punic Wars, don't fret. Just try to remember which civilization became a major world power and which civilizations practically have been forgotten. It shouldn't be too hard. After all, you have a bunch of questions about Rome, so that must mean the Romans became very powerful!**

F The Romans burned Carthage to the ground after the Third Punic War.

G The Punic Wars were fought between Rome and Carthage, not Rome and Athens. Athens had already faded by the time the Romans became powerful.

H The Romans were the undisputed power of the Mediterranean world after the Punic Wars. Their influence extended all the way around the Mediterranean Sea and all the way north to Britain.

J For a while, the Romans didn't have to share power with anybody. The Carthaginians were toast.

65 C **You don't need to know why education declined in the Middle Ages to answer this question correctly. If you can eliminate the three answer choices that list events that did not occur during the Middle Ages, you will answer this question correctly.**

A The rediscovery of classical Greek civilization occurred at the beginning of the Renaissance. It is one of the events that marked the end of the Middle Ages.

B The Christian Church was powerful during the Middle Ages, though it began to lose power after the Renaissance. The Church was also one of the few places where educated people could be found during the Middle Ages, so this is definitely not the answer.

C Yes. The fall of the Roman Empire ushered in the Middle Ages. Therefore, it's the only answer choice that makes any sense with the time frame. After the Roman Empire fell to the barbarians, education became much less important in a society dominated by small feudal kingdoms.

D No. The Middle Ages experienced a rise in limited monarchies, not absolute ones. The Roman Empire was an absolute monarchy, but during the Middle Ages, the power of the monarchy was subject to the power of the feudal lords.

66 J **Go with the most general answer choice.**

F Traditional societies, while not necessarily encouraging of rapid population growth, usually assist it by promoting marriage and family life and by discouraging family planning.

G Traditional societies usually emphasize religion.

H Traditional societies tend to teach virtues that pertain to an individual's responsibility to family and the community, thereby encouraging conformity as opposed to individuality. Order is usually very important in traditional societies.

J The word *traditional* is in direct contrast with the word *change*. Traditional societies thrive on order and predictability.

67 C **If you're guessing, remember to focus on the goals of the test. In other words, there must be something significant about the Neolithic Revolution, or the test writers wouldn't ask you about it. That said, pick an answer choice that describes something significant!**

A The Neolithic Revolution resulted in the cultivation of crops. There was a decrease in the number of nomadic tribes because many no longer needed to wander around in constant search of food.

B Prior to the Neolithic Revolution, tribes relied on hunting and gathering. The whole point of calling the Neolithic Revolution a *revolution* is that it changed things significantly. Namely, hunting and gathering was *out* and farming was *in*.

With farming, food surpluses were made possible. Food surpluses are essential to the establishment of towns and cities because in a society with a food surplus, not everybody has to be a farmer.

C Bingo. The Neolithic Revolution marked the beginnings of the development of villages and the establishment of organized governments.

D The Neolithic Revolution actually increased trade because agricultural surpluses and the creation of villages resulted in people having a lot more stuff to trade. Plus, with villages and town firmly established in permanent locations, regular trade routes could be established between them.

68 F **If you're familiar with West African civilizations, this question should be no problem. If you're not, think of what would be true of most societies so you can guess.**

F Every known civilization had or has some form of art, so this would be a good one to pick if you're guessing. Specifically in West Africa, clay sculptures were used extensively in religious ceremonies, and sometimes sculptures they were made from bronze and gold.

G West African civilizations did rely on trade—with each other! They also traded with civilizations in other parts of Africa.

H This was definitely not the case. West African civilizations were generally controlled locally, with various tribal leaders in charge of relatively small areas.

J West African civilizations emphasized the extended family or tribe.

69 D **The key to the correct response for this problem is determining why Japanese rulers were impressed. The reason was because St. Francis Xavier, a missionary who landed in Japan in 1549, impressed the Japanese by learning their culture.**

A Simply put, Europeans did not increase trade values in exchange for entry to Japanese.

B Enslavement was not associated with Asian exploration, nor was it threatened. The pioneers in East Asia began with the shipwreck of Portuguese in China. Their unruly behavior caused Chinese to describe them as "ocean devils." (It wasn't a very good start.)

C The change of monarchy had no specific impact on the impression conveyed during the exploration period. It is fair to say that diplomacy with any country relies on political leadership.

D Bingo. The reason was because of St. Francis Xavier, a missionary who landed in Japan in 1549. Rather than focusing on amassing wealth, St. Francis and the Jesuits attempted to accommodate themselves to Japanese culture. While the Japanese were not necessarily disposed to trade, some of the *daimyo* (military lords) found trade prospects appealing.

70 G Review the information in the Geography Skills section if you missed this question.

F Lines of latitude stretch east to west.

G All lines of longitude stretch from the North Pole to the South Pole, where the lines all meet. Look at a globe to see how they all come together, like slices of an orange.

H There are no lines that stretch from the equator to the prime meridian. The equator is a line of latitude, while the prime meridian is a line of longitude. To stretch lines between them would fill the globe with a series of diagonal lines.

J There are no specific lines on a globe that have the purpose of connecting Asia to the Americas. Some lines of latitude overlap both continents, but they extend all the way around the globe, stopping and starting on no particular continent.

71 D Think about these short quotations for a bit. A spark leads to a mighty flame. An acorn leads to a tall oak. One step leads to a thousand miles. Something small leads to something big in all three cases! What can we conclude from this?

A Don't take these quotes too literally. True, trees might make you think of conservation, and a journey of 1,000 miles might make you think of transportation, but all three quotes don't refer to transportation and conservation! Plus, avoid extreme language! We have only three quotes here, so there's no way we can conclude something about ALL cultures!

B None of these quotes have anything to do with making excuses. What's more, we can't conclude anything about ALL cultures from a mere three quotations.

C While geography is often an important factor in the development of a civilization, it doesn't have *anything* to do with these quotes. The quotes aren't talking about the impact of rivers, harbors, and good farm land!

D This is the best answer. We have three different people talking about different things (flames, trees, and journeys), but they're viewing them very similarly. The point is that you can achieve something big by starting with something small.

Chapter 9

Practice Test 2: World History from AD 1000 to the Present and Geography

What to Expect

The following practice exam contains 73 questions, which is the number of questions that are scored—63—on the real end-of-course World History from AD 1000 to the Present & World Geography SOL exam, plus 10 field-test items. **Remember:** The field-test items are the questions that are not scored but that you will have to answer anyway. You won't know which ones count and which one's don't, so do your best on every problem. Good luck!

Directions:

Read each question or statement carefully and then choose the letter of the answer choice that best answers the question or completes the statement.

1 **In European feudal society, an individual's social status was generally determined by —**

A birth

B education and training

C individual abilities

D marriage

2 **In Europe during the Middle Ages, increases in trade and commerce resulted in —**

F lower living standards for industrial workers

G decreased economic rivalry between kings

H increased political power for the clergy

J development of towns and cities

3 **The city of Jerusalem is important because it —**

A serves as a financial center of the Middle East

B is a major port for Israel

C has religious significance for Judaism, Christianity, and Islam

D has become the center of industrial development for Palestinian Arabs

4 **An important result of the Crusades in the Middle East was the —**

F increased tension between Muslims and Christians

G destruction of Muslim military power

H creation of a large Christian state on the Red Sea

J restoration of the Byzantine Empire

5 **The Magna Carta and English Bill of Rights are documents that —**

A limited the power of the monarchy

B established England as an independent state

C intensified the conflict between church and state

D decreased the wealth of the nobles

6 Many African nations changed their names after gaining independence. The Gold Coast became Ghana, Rhodesia became Zimbabwe, and the Belgian Congo became Zaire. These changes most closely reflect the idea of —

F nationalism

G pan-Africanism

H mercantilism

J capitalism

7 The Greek doctrine of Epicureanism differs from Cynicism because of the definition of —

A language

B love

C pleasure

D health

8 When President Wilson stepped on European soil in December 1918, huge crowds greeted him as the "King of Humanity" and "Prince of Peace," probably in reference to his —

F successful negotiations during the Cold War

G Fourteen Points plan for the world peace program

H victory against England in World War I

J ability to skillfully communicate in French

Use the map and your knowledge of world history to answer questions 9 and 10.

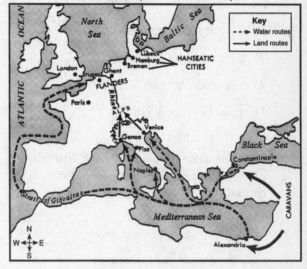

Trade Routes (13th – 15th centuries)

9 One reason that Italian city-states were able to dominate the trade pattern shown on the map was that they were —

A centrally located on the Mediterranean Sea

B situated north of the Alps

C unified by the Hanseatic League

D located on the trade routes of the North Sea

10 The development of trade routes along the routes shown on the map led to the —

F decline of the Greek city-states

G start of the Renaissance in Italy

H beginning of the Crusades to the Middle East

J first religious wars in Europe

11 "I will never allow my hands to be idle nor my soul to rest until I have broken the chains laid upon us by Spain."

This statement was most likely made by —

A a Latin American nationalist

B a Portuguese explorer

C a Roman Catholic bishop

D a Spanish conquistador

Use the map below and your knowledge of world history to answer questions 12 and 13.

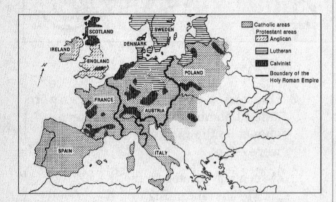

12 Which statement about the Holy Roman Empire is supported by the map?

F The religion of the people in the Holy Roman Empire was either Lutheran or Catholic.

G The Holy Roman Empire had fewer Protestant areas than the rest of Europe did.

H Calvinism was dominant throughout the Holy Roman Empire.

J Protestant influences were strongest in the northern areas of the Holy Roman Empire.

13 Which title would be the most appropriate for this map?

A "The Impact of the Protestant Reformation"

B "The Catholic Counter-Reformation"

C "The Fall of the Holy Roman Empire"

D "European Religious Unity"

The diagram below illustrates the social structure of feudal Japan.

14 The pyramid above shows that feudal Japan had —

F a classless society

G a growing middle class

H high social mobility

J a well-defined class system

15 Which statement is best supported by the existence of the African kingdoms of Songhai, Mali, Kush, and Nubia?

   A Natural geographic barriers prevented major cultural development in these civilizations.

   B Africans established thriving civilizations long before European colonization.

   C These societies were so involved with violent civil wars that there was little time for cultural development.

   D These African civilizations were entirely self-sufficient and discouraged trade with other areas.

16 The dominance of Christianity in Latin America and of Buddhism in Southeast Asia is a direct result of —

   F racial intolerance

   G cultural diffusion

   H urbanization

   J militarism

Use the graph and your knowledge of world history to answer question 17.

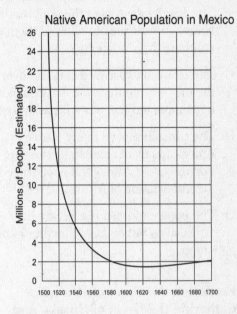

Native American Population in Mexico

17 Which statement is best supported by the information provided by this graph?

   A The Native American population in Mexico steadily increased between 1500 and 1700.

   B The Spanish conquest devastated the Native American population in Mexico between 1500 and 1540.

   C The Spanish conquest of Mexico improved the standard of living for the Native American population of Mexico.

   D Spanish influence in Mexico had ended by 1700.

18 Which statement best illustrates the contradictory actions of the Catholic Church in colonial Latin America?

F The Jesuits destroyed the temples of the Native Americans but allowed them to continue their religious rituals.

G The Church expressed concern over the mistreatment of Native Americans but supported the *encomienda* system.

H The Church moved many Native Americans from Spanish territory to Portuguese territory but encouraged the importation of African slaves.

J The pope endorsed the Treaty of Tordesillas but outlawed further exploration.

19 Frederick the Great of Prussia, Catherine the Great of Russia, and Joseph II of the Hapsburg Empire are all 18th century monarchs also described as —

A prominent entrepreneurs

B prominent writers

C devout religious figures

D benevolent despots

20 Judaism, Islam, and Christianity share a belief in —

F the central authority of the pope

G a prohibition on the consumption of pork

H reincarnation and the Four Noble Truths

J monotheism and ethical conduct

21 In the seventeenth and eighteenth centuries, Dutch interest in the islands of Southeast Asia was based mainly on the location's —

A rich deposits of gold and silver

B large numbers of Christian converts

C spice trade

D development of manufacturing potential

Use the speakers' statements below and your knowledge of social studies to answer questions 22 through 24.

Speaker One: I am offended by the term "Dark Continent." It implies that only ignorance and barbarism were here before European explorers and settlers invaded the continent, bringing their "enlightenment" ways.

Speaker Two: This hemisphere might have been a "New World" to Europeans, but it certainly was not a "New World" to the Incas, Aztecs, and other Indian nations whose worlds were destroyed by the brutal greed of the Europeans.

Speaker Three: Defeat at Dienbienphu resulted in French withdrawal from the region. I was glad to see them go. The French exploited our land, resources, and people. And now, American involvement in the region frustrates our nationalist ambitions. It will lead to more military conflict.

Speaker Four: Upon arrival, we found a primitive people living in a primitive land. We built roads and bridges, sanitation systems, schools, and hospitals. We helped eliminate starvation and poverty. I feel no guilt about our past and continued presence in the region.

22 Which speaker expresses an opinion that is different from the other three?

    F Speaker One

    G Speaker Two

    H Speaker Three

    J Speaker Four

23 The statement by Speaker One could best be used to support the idea that —

    A geographers have often disagreed on terminology

    B terminology and labeling can often lead to misconceptions and stereotyping

    C most native peoples welcomed the colonial experience

    D African economies are based on subsistence agriculture

24 The statements of all four speakers relate to —

    F socialism

    G isolationism

    H imperialism

    J feudalism

25 Deserts, mountains, and tropical rain forests have affected the history of many regions by —

    A encouraging foreign invasions

    B promoting political reform

    C isolating peoples and cultures

    D increasing social mobility

26 Karl Marx predicted that the overthrow of capitalism would follow a revolution with the —

    F proletariat overthrowing the bourgeoisie

    G colonists rising against the native population

    H traders demanding fancier goods

    J merger of military and democratic governments

27 Both the French and the British were interested in controlling Egypt in the mid-nineteenth century because of Egypt's —

    A control of the spice trade

    B industrial-based economy

    C vital mineral resources

    D strategic location

28 Enlightenment writers were primarily interested in —

    F changing the relationship between people and government

    G supporting the Divine Right Theory

    H debating the role of the Church in society

    J promoting increased powers for European monarchs

29 British control over South Africa, French control over Indochina, and Spanish control over Mexico are examples of —

    A isolationism

    B appeasement

    C nonalignment

    D imperialism

30 Which statement best explains why the Industrial Revolution began in Great Britain?

F The country had sufficient coal and iron ore reserves and good transportation.

G Industries were owned by the national government.

H A strong union movement was able to secure good working conditions for factory workers.

J Cities could accommodate the migration of people from rural to urban areas.

31 One similarity in the leadership of Peter the Great and Catherine the Great was that both —

A actively sought to westernize Russia

B instituted important reforms that gave citizens a voice in Russian government

C ended feudalism and improved the lives of Russian peasants

D supported ethnic nationalist movements within the Russian Empire

32 A primary cause of the French Revolution in 1789 was the —

F increasing dissatisfaction of the Third Estate

G rise to power of Napoleon Bonaparte

H actions of Prince Metternich

J execution of Louis XVI

33 By the eighteenth century, the mercantilist economic system was replaced by a laissez faire approach that essentially advised —

A *"Workers of the world, unite."*

B *"Let nature take its course."*

C *"The righteous shall inherit the earth."*

D *"No taxation without representation."*

34 The location of the Inca Empire in 1470 is the present location of —

F Ecuador, Chile, and Peru

G United States, Canada, and Mexico

H Brazil, Mexico, and Argentina

J Uruguay, Cuba, and Bahamas

35 "Revolution will occur more and more frequently in the industrialized nations as the proletariat struggles to overcome the abuses of the capitalist system."

This quotation reflects the idea of —

A Charles Darwin

B Karl Marx

C Niccolo Machiavelli

D John Locke

36 The reforms of the Meiji Restoration altered Japan's foreign and internal relations in all of the following ways except that they —

F abolished unequal treaties with foreign nations

G gained jurisdiction over foreigners on its soil

H abolished oligarchic rule and emperor worship

J promised citizens freedom from arbitrary arrest

37 A major result of the Opium War in China was —

    A an increase in the power of the emperor

    B the establishment of spheres of influence in China by Europeans

    C the expansion of Chinese influence to India and the Middle East

    D the expulsion of Europeans from China.

38 Many sixteenth century monarchs willingly accepted Martin Luther's rebellion against the Church because of all of the following reasons except that they could —

    F take control of property and revenues held by papal officials

    G strengthen political authority within their countries

    H increase citizen participation in government affairs

    J use Luther's supporters in their fight against Church authority

39 "With the new methods that used mass production, division of labor, the assembly line, and interchangeable parts, prices came down. A greater number of and variety of goods became available to more people. The domestic system was disappearing and a new revolution was sweeping across Europe."

Which revolution does this quotation describe?

    A French

    B Commercial

    C Russian

    D Industrial

40 As economically developing nations become more industrialized, which situation occurs most often?

    F The authority of religious leaders increases.

    G The traditional roles and values of women change.

    H The size of families increases.

    J The cost of medical care decreases.

41 Africa's geography is characterized by —

    A an absence of rain forests

    B an irregular coastline and excellent harbors

    C a diversity of topography

    D a lack of major river systems

42 The bedrock of ancient Indian civilization is best described by all of the following except —

    F reincarnation

    G Taoism

    H caste

    J karma

43 One major reason the Middle East is globally important today is because it —

    A represents a model of economic and political equality

    B allows European powers to retain spheres of influence

    C provides much of the petroleum used by industrial nations

    D remains a primary source of uranium

44 During the 1930s, the Nazi (National Socialist) Party of Adolf Hitler received support from the German people because it promised to —

   F abide by the Treaty of Versailles

   G improve economic conditions in Germany

   H promote policies that insured ethnic equality

   J utilize international organizations to bring peace

45 Which series of events is arranged in the correct chronological order?

   A The Treaty of Versailles is signed, then Adolf Hitler becomes chancellor of Germany, then German troops invade Poland.

   B German troops invade Poland, then The Treaty of Versailles is signed, then Adolf Hitler becomes chancellor of Germany.

   C Adolf Hitler becomes chancellor of Germany, then the Treaty of Versailles is signed, then German troops invade Poland.

   D The Treaty of Versailles is signed, then German troops invade Poland, then Adolf Hitler becomes chancellor of Germany.

46 Which term refers to the Jewish movement to establish a homeland in Palestine?

   F zionism

   G secularism

   H animism

   J Marxism

47 After World War II, which action was taken by many African territories?

   A demanding independence from their colonial rulers

   B refusing to join international organizations

   C rejecting most of the technology offered by Western nations

   D creating a strong, unified Africa

48 After World War II, the Soviet Union established satellites in Eastern Europe to —

   F support the remaining fascist governments in Eastern Europe

   G preserve capitalism in Eastern Europe

   H establish democratic governments in Eastern European nations

   J expand its power and control over Eastern Europe

49 Mohandas Gandhi is best known for his —

   A use of passive resistance to achieve Indian independence

   B desire to establish an Islamic nation

   C opposition to Hindus holding political office

   D encouragement of violence to end British rule

50 Which geographic factor has most strongly influenced Russia's foreign policies and economic development?

F lack of natural resources

G vast desert regions

H limited access to warm-water ports

J extensive mountain ranges

51 One characteristic of apartheid, which was practiced in South Africa, was —

A forced migration of blacks to other nations

B integration of all races in society

C an open immigration system

D segregation of the races

52 During the 1960s and 1970s, the primary reason for United States involvement in Southeast Asia was to —

F gain new markets for exports

G search for new sources of oil

H look for new colonies

J halt the spread of communism

53 The code of chivalry in Europe and the code of Bushido in Japan illustrate that —

A Different societies develop similar systems to meet similar needs.

B Vast societal differences exist between Eastern and Western cultures.

C Force is often used by nations to conquer other nations.

D The lower classes in society often want to be controlled by the upper classes.

54 The Sepoy Rebellion, the Boxer Rebellion, and the Mau Mau Uprising were reactions to —

F rapid industrialization

G European imperialism

H Mongol domination

J World War I

Use the bar graph below and your knowledge of recent history to answer questions 55 and 56.

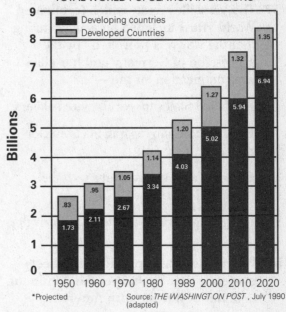

WHERE THE POPULATION GROWS FASTEST
TOTAL WORLD POPULATION IN BILLIONS

*Projected   Source: *THE WASHINGTON POST* , July 1990 (adapted)

55 Which statement is best supported by the data shown in the graph?

A The rate of world population growth has begun to decrease.

B The world's population tripled between 1970 and 1989.

C Most of the world's population lives in economically developing countries.

D The population of economically developed countries consumes most of the world's resources.

56 Which factor could be significant in explaining the difference in the growth rates between developing and developed countries?

F increased family planning in developed countries

G increasing pollution in developed countries

H the breakdown of extended families in developing countries

J the rise of single-parent families throughout the world

57 The Versailles treaty which terminated World War I included an article stating that the war was provoked "by the aggression of Germany and her allies" was justification for the —

A United States to recruit more soldiers

B Allies to claim reparations from Germany

C England to end relations with the United States

D Italy and France to end its tariff rivalry

58 Which statement provides the best evidence that Spain was the dominant colonial power in Latin America?

F Spain and Mexico continue to use the same currency.

G Spain continues to provide military support for Latin America.

H Spanish is the principal language spoken in most of Latin America.

J Argentina elects representatives to the legislature of Spain.

59 Which action would best help developing nations improve their standards of living?

A borrowing from the World Bank to purchase food for their citizens

B relying on a single cash crop for export sale in the world market

C encouraging an increase in the trade deficit

D investing in the development of human resources

60 Why are the Suez Canal, the Strait of Hormuz, the Dardenelles, and the Bosphorus considered strategic waterways?

F The nations that control these waterways have economic control over other nations.

G They are natural geographic boundaries and have often separated warring nations.

H They are located along the Tropic of Cancer, the Equator, or the Tropic of Capricorn.

J The nations that adjoin these waterways depend on them as a source of fresh water.

61 Which statement best describes an effect of geography on the development of Southeast Asia?

A The proximity of China promotes the growth of democracy.

B Large deposits of coal and diamonds attract Russian settlers.

C Vast areas of desert prevent exploration.

D The location of strategic waterways encourages trade.

62 Which geographic factor has had the most influence on Poland's historical and cultural development?

F a severe climate

G vast deposits of oil

H location on the Great European Plain

J a rugged coastline

63 Revolutions have most often occurred in nations in which —

A the majority of the people are economically prosperous

B social mobility is encouraged

C citizens can participate in the political process

D social, political, or economic dissatisfaction exists

64 Which statement describes the situation in Russia during the 200 years when the Mongols ruled?

F Russia experienced a cultural renaissance.

G Russia was isolated and paid tribute to the Khans.

H Westernization and industrialization began in Russia.

J Democratic reforms were encouraged in Russian society.

65 Which invention was fundamental to the Protestant Reformation?

A steam engine

B light bulb

C printing press

D gunpowder

66 The Native American population of Mexico in 1492 has been estimated at 25 million; the population in 1608 has been estimated at 1.7 million. This decrease was mainly a result of —

F crop failures brought on by poor weather conditions

G emigration of Native Americans to Europe and Africa

H wars between various native groups

J diseases introduced by the Spanish

67 During the Age of Absolutism (1600s and 1700s), European monarchies sought to —

A increase human rights for their citizens

B centralize political power in their nation

C develop better relations with Moslem rulers

D encourage the growth of cooperative farms

68 In the nineteenth century, the unification of Italy and the unification of Germany resulted in —

F upsetting the balance of power in Europe

G increasing competition for trade with Russia

H reducing feelings of nationalism in these nations

J encouraging a century of peaceful coexistence in Europe

69 Under Joseph Stalin, life in the Soviet Union was characterized by —

A an abundance of consumer goods

B political instability and numerous civil wars

C support for small family-run farms

D the use of censorship and the secret police

70 During the Cold War era (1945–1990), the United States and the Soviet Union were reluctant to become involved in direct military conflict mainly because of —

F the peacekeeping role of the United Nations

G pressure from nonaligned nations

H the potential for global nuclear destruction

J increased tensions in the Middle East

71 A major goal of the European Community is to —

A promote one-product economies

B forgive the debts owed to them by developing countries

C repay loans made by the United States to Western European nations

D make Western Europe economically competitive with Japan and the United States

72 One similarity between the Sepoys in India, the Boxers in China, and the Mau Mau in Kenya is that these groups —

F tried to drive Europeans out of their countries

G depended on Western support for their success

H adopted Marxist economic and political principles

J sought independence through nonviolence

73 Which document is an example of a primary source?

A a textbook on Russian history

B an encyclopedia article on religions of the Middle East

C a novel on the Age of Exploration

D the diary of a concentration camp survivor

Chapter 10

Answers and Explanations for Practice Test 2

Listed below are the answers to the practice test found in chapter 9.

1 A	11 A	21 C	31 A	41 C	51 D	61 D	71 D
2 J	12 J	22 J	32 F	42 G	52 J	62 H	72 F
3 C	13 A	23 B	33 B	43 C	53 A	63 D	73 D
4 F	14 J	24 H	34 F	44 G	54 G	64 G	
5 A	15 B	25 C	35 B	45 A	55 C	65 C	
6 F	16 G	26 F	36 H	46 F	56 F	66 J	
7 C	17 B	27 D	37 B	47 A	57 B	67 B	
8 G	18 G	28 F	38 H	48 J	58 H	68 F	
9 A	19 D	29 D	39 D	49 A	59 D	69 D	
10 G	20 J	30 F	40 G	50 H	60 F	70 H	

The field-test questions were numbers 6, 11, 27, 38, 43, 50, 57, 63, 66, and 72.

Learn From Your Mistakes

Review the explanations below for the questions that you missed, but don't just read the explanations! It's important that you also read the reason that the answer you picked is incorrect. If you understand why the wrong answers are wrong and why the right answer is right, you'll be less likely to make the same mistake when you take the real End-of-Course World History SOL exam.

What's more, try to figure out *why* you missed the questions that you missed. It's usually for one of two reasons: either you didn't understand the content of the question, or you were careless when reading the question and answer choices. If you didn't understand the content, review the relevant history and geography in the appropriate chapter in this book. If you were careless, slow down a little when taking the test and review the test strategies from chapters 1 and 2 of this book.

1 A **Feudal societies are traditional societies. Think of how an individual's social status is determined in a society that doesn't encourage education, doesn't permit social mobility, and provides few avenues for people to consider doing something other than what their parents do.**

A This is the best answer. Birth determined a person's status. If a person was born to parents who were peasants, he became a peasant.

B Education and training were not vehicles for social mobility. Only the ruling class was educated. The peasants received one-on-one training from their parents so that they could perform the same jobs.

C Abilities were not related to social status. Peasants had few opportunities to use their talents or abilities. And even if they did, this did not improve their social standings.

D Peasants married peasants. Rulers married family members of other rulers. Enough said.

2 J **The correct answer to this question is the most general of the four answer choices and the hardest one with which to argue. Ask yourself: In what would trade and commerce most likely result?**

F If you realize that the Industrial Revolution didn't occur until the mid-eighteenth century, and that the Middle Ages ended 300 years earlier, then you've got to conclude that whatever happened in the Middle Ages probably didn't really affect the living standards of industrial workers.

G In reality, kings during the Middle Ages didn't trouble themselves too much with economic rivalries, except in the broad sense when land was involved. However, even if you didn't know this and thought that kings were interested in economic rivalry, then it makes sense that increases in trade and commerce would result in *increased*, not decreased, economic rivalries. So this answer wouldn't make sense anyway.

H Increased trade and commerce decreased the power of the clergy in two ways: People became concerned with the material as opposed to the spiritual, and since trade exposed people to ideas from different parts of Europe and the world, people became influenced by a world that was much larger than their local clergy were prepared to respond to.

J When people want to buy and sell, they go to where the people and stores are. It simply makes sense that increases in trade and commerce will aid the development of cities.

3 C If you have reviewed your major world religions, you will probably get this question right. If not, try to eliminate answer choices that you know are not correct.

A No. There is no single, major financial center of the Middle East, but if one would be named, it wouldn't be Jerusalem. Tel Aviv, Tehran, and Beruit are more significant financial centers.

B Know your geography. Jerusalem isn't on the sea.

C Yes! Jerusalem is significant to all three religions.

D There is no center of industrial development for Palestinian Arabs.

4 F Choose the answer choice that is most difficult to argue with. In other words, choose the answer choice that you are certain has occurred, even if you're not sure why it occurred.

F There definitely has been tension between Muslims and Christians, and if you know anything about the Crusades, you would understand why. The Crusades were led by European Christians who intended to win back the Holy Land (Palestine) and convert Muslims to Christianity. When it turned out that Muslims were just as secure in their religion as Christians were in theirs, tensions rose. The Muslims maintained control of the Holy Land through more than a century of battles.

G No. The Muslims retained control of the Holy Land.

H No Christian state was established on the Red Sea. No permanent Christian state was established anywhere else in the Holy Land as a result of the Crusades.

J This answer choice doesn't make a lot of sense because it implies that the Byzantine Empire had been defeated and therefore needed to be "restored." In fact, the Byzantine Empire had not yet been defeated during the time of the Crusades, though it was beginning to falter. In any case, the Crusades did not "restore" the Byzantine Empire.

5 A If you're familiar with these two documents, this question shouldn't be a problem. If not, you should learn about them before taking the test. In the meantime, choose the answer choice that best describes a general result of individual people gaining rights.

A Yes. The Magna Carta and English Bill of Rights limited the power of the monarchy. King John signed the Magna Carta in 1215 under pressure from the noblemen, who gained due process rights and tax rights. The English Bill of Rights was signed centuries later and limited the power of the monarch even more. This answer should make sense to you since any time citizens gain rights, the government cannot legally do anything to abridge those rights.

B England was an independent nation long before the Magna Carta and the English Bill of Rights.

C The documents did not directly intensify the conflict between church and state. The documents helped to resolve problems between the monarchy and the nobles/citizens, not between the church and the state.

D The nobles gained wealth as a result of the documents, which is one reason they wanted them signed so badly.

6 F The key phrase in this question is "gaining independence." The question gives you several examples, but you don't even need them. If you don't know the words in the answer choices, look them up now. Which answer choice reflects the desire to change the name of your country after winning independence?

F Nationalism is a motivating force behind an independence movement in the first place. In these situations, the native people wanted to reassert their own culture and wipe away the remnants of colonization. Changing the names from British names to African names was a way of reclaiming their African heritage.

G The name changes were not the result of Pan-Africanism. The name changes occurred independently by each nation. *Pan-Africanism* is a policy whereby the countries of Africa work together for common goals.

H The independence movements wiped the policy of mercantilism from the face of Africa. Mercantilism is closely associated with colonialism.

J The name changes have nothing to do with economics, or with an economic philosophy, whether it be capitalism or something else.

7 C Epicureanism is the Greek doctrine defined as the art of making life happy, with pleasure. On the other hand, Cynicism, another Greek doctrine, proposes that the only good is virtue.

A Language is an inappropriate answer because neither philosophy addresses language.

B No. This answer choice may be romantic, but it's incorrect. Either philosophy may eventually lead to some form of love, but love is not the essence of these philosophies.

C Yes, pleasure is the difference between the two definitions. Epicureanism is the Greek doctrine defined as the art of improving life with pleasure. Pleasure, not indulgence, is further defined as serenity and avoidance of pain. Remember the saying: "Eat, drink, and be merry for tomorrow may never come." More likely, that's an Epicurean speaking. According to Epicureanism, pleasure is the highest and *only good* reason for living. On the other hand, Cynicism, another Greek doctrine, proposes that the *only good* is virtue. Everything else, including riches, honor, freedom, even life itself, is contemptible.

D Health may be a byproduct of Epicureanism, uʃî it is a conspicuously incorrect answer.

8 G Check your time line for 1918, and remember this is the post-World War I period. Use that knowledge to eliminate a few wrong answers and find the right one.

F This answer choice is from the wrong time period. It's an incorrect response primarily because the Cold War was the aftermath of World War II. Wilson was the United States President during World War I. This should have been an easy cross-off using POE.

G This is precisely correct. Wilson's Fourteen Points stirred hopes for permanent peace after World War I. Europeans believed Wilson's plan to be workable and it offered hope for rebuilding the devastation left in Europe. A significant component of Wilson's plan was national self-determination. The defeated Germans thought that the Fourteen Points offered the best chance for restoration of peace.

H The United States and England were victorious allies in World War I. Consequently, England and the United States were not at war, and Wilson certainly didn't win victory over England.

J President Wilson's French was unrelated to his warm European welcome.

9 A **If you know your geography, this question is easy. Just eliminate answer choices that are way off the target. If you have no idea where Italy is on the map, you're going to have some problems. Still, try to think of the answer choice that describes a location that would dominate a trade pattern.**

A This is it. Italy is the long peninsula that is sticking out into the Mediterranean Sea. The Italian city-states included Venice, Genoa, Pisa, and Naples. It's centrally located on the map! Even if you weren't sure which part of the map is Italy, you should still guess this answer because it just makes sense that a central location would dominate trade patterns.

B Italy is not situated north of the Alps. Flanders is.

C Lubeck, Hamburg, and Bremen were unified by the Hanseatic League, but none of them are in Italy!

D This map doesn't give us any trade routes through the North Sea, so definitely eliminate this one. (There *were* trade routes in the North Sea; however, it's just not the point of *this* map).

10 G **Look at the map carefully. The title informs us that the time period is from the thirteenth to the fifteenth centuries. The arrows on the dotted lines and solid lines inform us that stuff is moving from the Middle East and Africa to Europe (almost all the lines are moving north and northwest). Therefore, choose the answer that describes a situation that would have resulted from the movement of this stuff.**

F No. The Greek city-states declined way before the time period of this map. This map shows us Italian city-states, not Greek city-states (like Sparta and Corinth).

G This is the best answer. You should be aware that the Renaissance occurred in the fifteenth and sixteenth centuries, so it simply makes sense that the centuries just prior to the Renaissance would have led to the Renaissance. Also, you should be aware that the Renaissance was, in part, a rediscovery of the culture of the ancient world. The lines on the map show information and goods flowing to Europe from places like Alexandria and Constantinople (the holding places for many of the ancient documents and ideas). It was Europe's increased communication with these places that enabled it to rediscover its lost past.

H No. The Crusades were during the eleventh, twelfth, and thirteenth centuries. In other words, the Crusades are significant to developments that led to this map, not the other way around. Even if you're not sure of the dates of the Crusades, you should be aware that the Crusades were launched from Europe into the Middle East, in which case the arrows on the map would have to pointing in the other direction.

J No. Religious wars in Europe occurred well before the thirteenth century.

11 A **Focus on the words "broken the chains laid upon us by Spain." You want to find a person that would be angry about Spanish control.**

A Latin American nationalists fought against Spanish colonialism and succeeded in winning independence, thereby breaking the chains laid upon them by Spain.

B Portuguese explorers would not have had chains laid upon them by Spain. They had nothing to do with Spain. They're Portuguese!

C Spain promoted the expansion of Catholicism into Latin America. Roman Catholic bishops were not enchained by Spain.

D Spanish conquistadors conquered Latin America for Spain. They were rewarded by Spain, not enchained by Spain.

12 J **Look at the map closely and study the key. Now, start eliminating answer choices.**

F If this were true, then why does the map show that there were Calvinist areas?

G How big is an area? This answer choice is too ambiguous. We can count people, but we can't count areas unless we know what is meant by the word *area*.

H Just by looking at the map, we can tell that a lot more space is non-Calvinist than Calvinist.

J In the south, there are a lot of dots (Catholic areas). In the north, there are a lot more stripes and dashes (Protestant areas). Looks like we have ourselves an answer!

13 A **The map clearly contrasts Catholicism with Protestantism. That's why the mapmakers shaded areas differently: so that you tell which regions had which types of Christians. Protestantism developed after Catholicism. The areas on the map that are shown as Protestant were formerly Catholic. What's the best title for this map?**

A Yes! That's why the key on the map makes a distinction between Catholic areas and Protestant areas: so you can see that the Protestant Reformation had an impact on large areas of Europe.

B This map doesn't tell us anything about the Catholic response to the Protestant Reformation.

C This map doesn't tell us anything about the loss of territory. It only tells us the dominant religions in certain areas during a specific time.

D Absolutely not! If there were unity, the entire map would be covered with dots or the same kind of stripe.

14 J Whenever pyramids are used, it means that the important or powerful people are at the top (and that there are fewer of them) and that the less important and less powerful people are at the bottom (and that there are more of them).

F There are definitely classes. That's why there are really dark lines separating the labels (and why there are labels in the first place).

G There wasn't a significant middle class in feudal Japan. Even if one of these classes were a middle class, though, we wouldn't know if it was growing or diminishing unless we knew how big it was during some previous time.

H This pyramid doesn't tell us if people were actually able to move up the pyramid. In fact, they were not able to do so, but this pyramid does not illustrate that fact.

J There are four groups. There are lines in between them. Groups are ordered from top to bottom. Bingo!

15 B You need to know about only one of these civilizations to answer this question correctly. If you don't know about any of them, you should because they're all fascinating civilizations. In the meantime, don't fret. Instead, pick an answer that is consistent with the goals of the test writers.

A All of these civilizations had major cultural development. Do you really think the test writers would bother asking about civilizations that weren't significant?

B This is the best answer. All of these civilizations thrived long before the Europeans colonized the African continent. One of the goals of the test writers is to instill an appreciation for cultures apart from those in Europe and North America. This answer choice screams, "Pick me!"

C All of these civilization nurtured significant cultural development. If they hadn't, why would the test-writers be asking about them?

D This answer choice uses extreme words ("entirely self-sufficient") and therefore is an unlikely choice even if you're not sure of the trading practices of these civilizations. In fact, these civilizations grew wealthy due to trade with the civilizations of the Middle East and beyond.

16 G You need to know the basics of Christianity and Buddhism, as well as the basics of all major religions, because they're all over the test. But to answer this question correctly, you only need to know about either one of them. You should also be aware that only one of the words in the answer choices makes sense with regard to the spread of religion in certain regions of the world.

F We want an answer choice that describes the direct cause of the dominance of religions in certain regions of the world. While religious beliefs are sometimes used to justify racial intolerance against certain groups, racial intolerance is not a reason that these two religions *spread* to different groups.

G This is by far the best answer. Why are these religions dominant in places where these religions did not originate? Because these religions spread to new areas. How did they spread? Through cultural diffusion. Christianity started in the Middle East, gained steam in Europe, then spread to Latin America through cultural diffusion associated with European colonization of the region. Buddhism started in Northern India and spread centuries later to Southeast Asia when Ashoka, a powerful emperor who extended his influence throughout the region, sent missionaries into the surrounding regions. Later, Buddhism spread along the trade routes between India, China, and Southeast Asia as merchants took not only their goods on their travels but also their religion.

H Christianity and Buddhism are practiced in rural communities as well as in urban communities. If anything, urbanization has led to a decline in the influence of Christianity and Buddhism.

J While Christianity was enforced to some degree in Latin America by European militarists, it also spread to some degree by those who did not use force and who even objected to the use of force. And Buddhism definitely can't be classified as a militaristic religion. It generally spread via missionaries and cultural diffusion.

17 B **Look at the graph carefully. The horizontal lines represent millions of people (2 million, 4 million, etc.) The vertical lines represent 20-year increments (1500 through 1700). The graph is titled "Native American Population in Mexico," so the black line represents the number of Native American living in Mexico from the year 1500 through the year 1700. Notice that the black line drops sharply (which means that the population of Indians is dropping sharply) during the first few decades of the period. What can we conclude with certainty?**

A No! The population decreased, not increased. If it increased, the black line would be rising as time passed, not falling.

B According to the graph, the most severe drop in the Native American population of Mexico occurred from 1500 to about 1540. In 1500, the population was somewhere near 26 million. By 1540, it dropped to about 6 million. That's a loss of 20 million Native Americans in 40 years. That's pretty severe!!

C This graph doesn't tell us anything about the standard of living of the Native Americans population; it tells us only the number of Native Americans were living in the first place. As for the 2 million Native Americans that were still living in Mexico in 1700, we have no idea, based on the information in the graph, whether their standard of living increased. Just in case you're wondering: the standard of living of the Native Americans didn't improve in any significant way during this time period.

D First of all, this graph gives us information only through 1700, so we have no idea what happened after 1700. Second, we don't know if what did or didn't happen after 1700 was due to the influence of the Spanish. Third, you should already be aware that the Spanish influence didn't end by 1700. To this day, the Spanish influence can most readily be seen in Mexico's language (Spanish), dominant religion (Catholicism, brought by the Spanish), and architecture.

18 G Try to think of everything that you know about the Catholic Church, specifically in Latin America. Then focus on the words of the question. The correct answer must state two things that (a) are contradictory and (b) actually occurred. Eliminate answer choices that either don't state a contradiction or never even happened in the first place.

F The Jesuits tried to convert the Native Americans. They didn't allow them to continue their religious rituals. Therefore, this answer choice does, in fact, state a contraction, but the second half never happened.

G This is the best answer. The church was ideologically tied (and indebted) to the political and practical justifications for imperialism and, therefore, supported the *encomienda* system, which ultimately enslaved much of the Native American population in colonial Latin America. Yet, despite the fact that the Church had little respect for Native American cultures and beliefs, many people in the church believed it was their moral duty to protect their fellow human beings from blatant mistreatment that resulted from the *encomienda* system. So there was quite a paradox indeed! If the church truly wanted to help the Native Americans, it would have opposed the *encomienda* system entirely. Instead, it supported the institution that caused the problem and then tried to argue against the consequences of the problem. Incidentally, this whole mess continues to be played out within the Catholic Church in Latin America. Even today, conservative Catholics in Latin America remain very loyal to the political policies of the leadership, while more liberal Catholics insist that the church should take a much more active role in helping the victims of those same political policies.

H The church didn't move Native Americans from Spanish to Portuguese territory, although the church did play a major role in drawing the lines between Spanish and Portuguese territory in the first place. The church also didn't encourage the importation of African slaves, but it also didn't make any significant attempt to stop it.

J This answer choice doesn't give us a contradiction. The Treaty of Tordesillas was an agreement between Spain and Portugal in 1494 regarding the division of Latin America between them. Nothing about a boundary agreement between two nations is contradictory with the outlaw of further exploration. Of course, you might already know (or have reasoned) that further exploration wasn't outlawed. Therefore, this answer choice is wrong for two reasons. Incidentally, the pope did sign a declaration dividing the New World one year before the Treaty of Tordesillas, but the Treaty between the two nations moved the pope's lines around.

19 D This question lists three leaders representing three different countries during the 18th century. What is the one similarity among these three?

A While all three impacted commerce in their separate countries, entrepreneurship was not the outstanding similarity of the three.

B Frederick the Great of Prussia, Catherine the Great of Russia, and Joseph II of the Hapsburg Empire were not prominent writers.

C This is a wrong answer. These three leaders were certainly *not* religious figures. In fact, clergy and aristocrats were relentless in objecting to many of these monarchs' reforms.

D Yep, they were all benevolent despots who believed that the divine right of monarchs should be used for the benefit of people. Catherine frequently applied slogans of the Enlightenment such as "All citizens ought to be equal before the law" or "Sovereigns are made to serve their people." Frederick advanced agricultural reforms in Prussia. Joseph was the most sincere because he attempted to remold his empire in accordance with his reform principles.

20 J **This is a good POE question. Start getting rid of answer choices that don't apply to even one or two of the religions in the question.**

F Only some Christians, namely Catholics, believe in the central authority of the pope. Non-Catholic Christians, Jews, and Muslims don't.

G Judaism and Islam prohibit the consumption of pork, but Christianity does not.

H None of these religions espouse reincarnation or the Four Noble Truths. These are tenets of Buddhism.

J Strong systems of ethical conduct and the belief in a single supreme being are central to all three religions.

21 C **If you just plain don't know the history of mercantilism in Southeast Asia, this one might be a bit difficult, but hang in there. First, eliminate answer choices that likely don't make sense with the time period of seventeenth and eighteenth centuries (or, put another way, 1600s and 1700s). Second, recall that generally speaking, European colonization was driven mostly by economic greed and conquest, and secondarily by religious motivations. You might have to guess, but hopefully you'll remember which region of the world has traditionally been known for its spices and which region of the world has traditionally been known for its gold and silver deposits.**

A The islands of Southeast Asia are not known for rich deposits of gold and silver. Europeans were very interested in South Africa and South America for these two minerals.

B The Dutch were not primarily motivated by religion when it came to colonization of Asia. The Spanish were a bit more motivated by religion, although they were primarily motivated by economic concerns.

C This is the best answer. The Dutch were primarily interested in the islands of Southeast Asia because of the spice trade. They established colonies in the East Indies (present-day Indonesia). They set up trading posts in their colonies, competing not only with other European powers, but also with Arab, Indian, and Chinese merchants for the largest share of the spice trade.

D We're talking here about the 1600s and 1700s. The Industrial Revolution, however, didn't gain steam until the late 1700s and into the 1800s. Therefore, this answer choice is out of the time period. What's more, even after Europe industrialized, it wasn't very interested in the islands of Southeast Asia as potential manufacturing sites. Instead, it looked to Southeast Asia to provide it with raw materials that were sent back to the factories in Europe. This was all part of the process of mercantilism, which you need to look up in the review materials if you've forgotten what it is.

22 J Read the quotations carefully. You should be able to determine that Speaker One believes that it is incorrect to claim that the Europeans enlightened Africa. Speaker Two mocks the use of the term "New World" by the Europeans, since North and South America were not new at all to the natives that were overrun by the greed of the Europeans. Speaker Three is frustrated with the French and then American interference in Vietnam. Only Speaker Four attempts to justify colonialism. Which one of these is different from the others?

F Speaker One expresses frustration with colonialism, just like Two and Three.

G Two is similar to One and Three.

H Three is angry just like One and Two.

J Only Speaker Four attempts to justify colonialism.

23 B Speaker One believes that it is incorrect to claim that Europeans enlightened Africa. The speaker believes that the use of the term Dark Continent implies that the continent of Africa was uncivilized and unsuccessful before the arrival of the Europeans.

A The term *Dark Continent* was not used by geographers for Africa. Geographers simply used the term *Africa*. Instead, *Dark Continent* was a term used by colonialists who wanted to justify their involvement on the continent.

B Yes! The speaker is frustrated with the use of the term *Dark Continent* because it only serves to reinforce untrue and unfair stereotypes.

C No. The speaker would be angry at the suggestion that the colonialists were welcome.

D This quote has nothing to do with the agricultural systems of Africa.

24 H All four speakers are referring to the involvement of European powers in other parts of the world. What word best describes this kind of involvement?

F Socialism is an internal economic policy that many nations have adopted, but it does not describe involvement in other nations.

G Isolationism is the practice of noninvolvement in the affairs of other nations and is the opposite of what is being described by the speakers.

H Yes! Imperialism is colonialism. It is the practice of building an empire by establishing colonies.

J Feudalism is entirely unrelated to this question. It is a social system composed of small manors ruled by feudal lords. The speakers, however, are referring to centralized empires that are colonizing the rest of the world.

25 C **Think of a sweltering, expansive desert, mountain peaks so tall and steep that it hurts your neck to look up at them, and tropical rain forests so thick and full of humidity that your skin is covered in nothing but sweat and bugs. Now, how are these places going to affect the development of a region?**

A Who wants to invade those places? And even if they do, how are they going to get through the desert or over the mountains?

B Nothing about a mountain or a rain forest screams "democracy" or "theocracy." Absolutely nothing. Political reform occurs in all sorts of places, but it's for political reasons, not usually geographic reasons.

C Extreme heat. Mudslides and avalanches. Three straight months of rain and weeds. These are the things that tend to isolate people, keeping them from moving away and keeping others from moving in.

D Isolated places tend to remain unchanged for long periods of time.

26 F **You need to know as much about Karl Marx as possible for the exam. If you knew about him, perhaps you knew that Karl Marx predicted that the overthrow of capitalism would follow a revolution with the proletariat overthrowing the bourgeoisie.**

F Karl Marx, a proponent of the class struggle theory, was convinced from his history studies that capitalism would be overthrown by a class struggle between the bourgeoisie (industrialists and professionals) and the proletariat (the working class in factories, mines, and other industrial facilities). Marx foresaw a proletariat victory which would overthrow capitalism.

G This answer choice is incorrect because the identified revolution players are inaccurate. Marx anticipated uprisings in the colonies, but *not* against indigenous populations. Instead, he predicted that profitable colonies, similar to their mother country, would become industrialized capitalist colonies. Because the old manufacturing centers of the West would eventually decline, leaving workers with high unemployment, the ensuing revolt would occur between the unemployed western workers and the wealthy capitalist colonies.

H This is an implausible response because traders would unlikely revolt for fancier goods. Perhaps they would revolt against higher prices, but not fancier goods.

J Marx was unconcerned about a merger between military and democratic governments.

27 D **Eliminate answer choices that do not apply to Egypt. If you're left with more than one answer, choose the answer choice that best describes why two very powerful nations with plenty of other colonies would be particularly interested in Egypt.**

A Egypt doesn't produce spices. India does, however, and that was a reason for British colonization there.

B Egypt did not have an industrial-based economy. Besides, Britain and France already had industrial-based economies. They were interested in Egypt because it would give them something that they needed.

C Egypt does not have vital mineral resources. What resources it does have Britain and France could attain from their other colonies.

D Yes! Egypt is located in northeastern Africa and is bounded by the Mediterranean and Red Seas. It was the ideal location for the building of a canal that would link the two seas. The canal, later built by a French company and known as the Suez Canal, was important because it eliminated the need for ships to travel around the tip of Africa in their journeys between Europe and Asia, thus making the trips much shorter.

28 F **Just like the Reformation and the Renaissance, the Enlightenment was all about change. Think of it this way: If you need to put light on something, it was in the dark. Therefore, the writers of the Enlightenment believed that they were taking something out of the dark and putting it into the light; they believed they were producing change.**

F The relationship between people and their government is precisely what the writers of the Enlightenment wanted to change. The desired change was in response to the Age of Absolutism wherein monarchs thought they had a divine right to rule without regard to the rights of the citizenry.

G This wouldn't have changed things. It would have kept things the same. Therefore, it wouldn't have been *enlightening*.

H Many Enlightenment writers did question the role of the Church in society, but they did not debate it with each other because almost all of them agreed that the Church was too powerful and extended itself beyond religious matters.

J Absolutely not. Just like answer choice **G**, this answer choice represents the opposite of what the Enlightenment writers wanted to accomplish. They wanted the citizenry to understand their own rights and to assert those rights, thereby limiting the powers of the monarchs.

29 D **If you don't know the definitions of the words in the answer choices, you're not going to be able to guess very well. Which word is related to one country controlling areas outside of its own boundaries? Which word is often used in reference to countries establishing empires?**

A Isolationism is the practice of staying uninvolved in things outside of your own territory. If the British kept to themselves, they certainly wouldn't have had control over South Africa.

B Appeasement means giving into the requests or demands of somebody else so that you can avoid getting into a fight with them. What does French control over Indochina have to do with avoiding fights? Absolutely nothing. That's why it's not the right answer.

C Nonalignment is a position taken by a group that doesn't want to take sides in a battle between two other groups. Spanish control over Mexico only involves two countries, with Spain in control. Since there aren't three or more countries involved in the matter, nonalignment isn't even a possibility.

D Imperialism is the practice of acquiring control of someone else's territory or economy or institutions. Clearly, it's the only word that relates to the scenarios in the question. Plus, it's an extremely important word, which is why there is so much discussion regarding imperialism in the review materials.

30 F If you don't know what the Industrial Revolution is, look it up in the review materials! Okay, now let's think about this question for a minute. It's asking you to determine why the Industrial Revolution began in Great Britain. Therefore, the right answer has to be something that was true of Great Britain before the establishment of large industries, or else those industries would not have developed in Great Britain in the first place. As such, you want to pick an answer choice that describes something that would attract the development of industry, as opposed to an answer choice that merely describes something that results from the building of industry. In other words, you're looking for a cause of the Industrial Revolution in Great Britain, not an effect of the Industrial Revolution in Great Britain.

F This is the best answer. The Industrial Revolution was sparked by the invention of the steam engine, and coal is needed for the steam engine to work. The steam engine was the catalyst of the factory system, which needed good systems for transporting raw materials to the factories and finished products away from the factories. The Industrial Revolution itself also improved transportation systems dramatically through the creation of the steam-powered locomotive, which, of course, ran on iron rails. But even before the development of the railroad, Great Britain had relatively good transportation systems in place.

G In fact, industries in Great Britain were generally private enterprises. But even still, you should have eliminated this answer choice because the question is asking you to locate something that would result in the development of industries in the first place, not something that describes who owned the industries once they were developed!

H The development of unions occurred after the Industrial Revolution resulted in hundreds of thousands of low-paying and dangerous jobs in the factory system. This can't be the answer because the question is asking you to find a *cause* of the development of the factory system in Great Britain, not an *effect* of the factory system.

J Actually, cities were overwhelmed with the influx of new residents during the Industrial Revolution. Multiple families crammed into single residences. City services were severely strained. But you don't have to know about the development of cities to eliminate this answer choice. You only have to wonder about one thing: Why would British cities be different from cities in other countries in terms of the ability to accommodate new immigrants? Almost all cities are bustling and crowded. That's what makes them cities.

31 A You only need to know about either Peter the Great or Catherine the Great to answer this question correctly. Start eliminating answer choices that you know don't apply to either or to both of them.

A Yes. When you see either name, you should automatically think "Westernization." Both Peter the Great and Catherine the Great actively sought to westernize Russia's industry, economy, customs, and attitudes. Peter the Great even went so far as to build the new capital of St. Petersburg, which was in a more western location than the previous capital of Moscow.

B Neither Peter nor Catherine westernized Russia's government. Instead, both ruled as near-absolute monarchs, sometimes viciously. They did allow feudal lords to control their own lands, but the citizens themselves had no voice in the Russian government.

C No. Neither Peter nor Catherine ended feudalism. The lives of the peasants were not improved because the feudal lords continued to exploit them.

D No. Neither Peter nor Catherine supported ethnic movements. Instead, they actively crushed the movements in an effort to keep Russia united under their rule.

32 F **The French Revolution is discussed extensively in this book, so make sure you review it. This question is quite detailed, however, so even if you remember the basics of the French Revolution, you might not remember all the people in the answer choices. As always, when you're guessing, it's important to pick an answer choice that describes something general and hard to argue with because it's more likely that something general will be correct than something too specific.**

F Why do revolutions occur? Because people are dissatisfied, for one reason or another. All over the world at different points throughout history, revolutions have occurred because a large chunk of the population was frustrated with the status quo. The Third Estate was a term given to the middle class and peasants. It was their dissatisfaction that led to the revolution. Refresh your memory by reviewing the French Revolution in this book.

G Napoleon rose to power as a *result* of the French Revolution. Review Napoleon in this book, especially if you thought this was the right answer.

H Metternich was an Austrian Prince who had nothing to do with the cause of the French Revolution.

J The execution of Louis XVI occurred well after the revolution began. Louis XVI's execution was a consequence of the revolution, not a cause of it.

33 B **Laissez faire is a very important term to know, so if you don't know it, now is the time to go back and review. If it helps you remember, try to see the word "lazy" in "laissez" and realize that laissez faire is the term given for "lazy" government—one that lets people govern themselves. Answer Choice B works best for that.**

A Wrong philosophy and wrong time. *"Workers of the world, unite"* emerges from the last sentence of Karl Marx's 1848 *Communist Manifesto*. Marx pled for brotherhood and cohesion among the world's workers.

B This answer choice is correct because *laissez faire* encouraged individualism in direct opposition to mercantilism's rigid regulation of economic life. *Laissez faire* economics essentially advised to let the people do what they will. Adam Smith substantiated *laissez faire* in his book, *Wealth of Nations* (1776), with the argument that individuals are motivated by self-interest and they do as they please for personal objectives.

C This response is unrelated to mercantilism or *laissez faire*. It is relative to the spirit of Christianity in its appeal to the poor and humble.

D This quote comes from the American colonies' fight for independence, not for *laissez faire* economics.

34 F **This is a straight geography question. Use whatever you know about the Incas Empire to help you solve it. If you aren't sure of the right answer, try throwing out any choices that you know for certain are wrong.**

F Located in South America along the Pacific Ocean, the Inca Empire stretched along the Andes Mountain range that borders the Pacific Ocean. The present borders are Ecuador and Colombia in the north to central Chile in the south, with Peru midway. The population of the Inca Empire numbered about 12,000,000 with an estimated 100 ethnic groups speaking 20 unrelated languages. The arrival of Pizarro in 1532 started the collapse of Inca power, followed by decimation of the population from the strain of foreign diseases.

G Oops, this is the wrong answer because the Inca Empire was based in the South rather than North or Central Americas, which is the location of the United States, Canada, and Mexico.

H This is the wrong location again, because both Brazil and Argentina bordered the Inca Empire, but are closer to the Atlantic rather than the Pacific Ocean.

J Uruguay, Cuba, and Bahamas are all proximate to Atlantic Ocean and its tributaries. Uruguay is located farther south in South America on the Atlantic Ocean.

35 B **This person doesn't like capitalism. The person predicts the downfall of capitalism. Do you know the folks in the answer choices? If not, look them up now.**

A Darwin was a scientist, not a philosopher or economist.

B Yes! Karl Marx was the leading communist theorist, and he advocated communism as a direct alternative to the abuses of capitalism. (Anytime you see the word *proletariat* in a question, you can pretty much guess the correct answer will have something to do with Marx!)

C Machiavelli was a political philosopher, but this quote certainly wouldn't have come out of his lips. He authored *The Prince*, which was written to instruct leaders to maintain their power by achieving their political goals through any means necessary.

D Locke was a political philosopher, not an economic one. While he would agree that people are destined to overcome abuses by their leaders, he was not anticapitalist. He simply believed in responsible government.

36 H **The Meiji Restoration was an important step in modernizing Japan, primarily by adopting Western standards. The only answer choice that listed something that the Mejii Restoration didn't do was H, abolish oligarchic rule and emperor worship.**

F An important reform of the Meiji Restoration was abolishing unfair foreign treaties. Based on the argument that Japan ranked among the civilized nations, they demanded national sovereignty.

G By 1899, Japan had gained legal jurisdiction over all foreigners on its soil. Consequently, Japan became the first nation to break the chains of Western control.

H This is the correct answer choice. Constitution reforms were indeed adopted along with a form of parliamentary government. The first article of the constitution, however, provided that "The Empire of Japan shall be reigned over and governed by a line of Emperors unbroken for ages eternal." The third article described the emperor as sacred and inviolable. What remains? Constitutional government under an emperor.

J Constitutional reforms protected citizens from arbitrary arrest in addition to protection of property, freedom of religion, and association.

37 B **The Opium War was bad news for China. Britain wanted to trade with China. When a British ship was stopped by the Chinese in Canton, Britain attacked China and forced the emperor to sign treaties allowing Britain trading rights. Things only went downhill from there for China. Other European nations also wanted a piece of the huge Chinese pie and carved China up into spheres of influence. It was humiliating. It was degrading. It was European imperialism.**

A No. The emperor and China were weak, not strong.

B Yes. The Opium War was the first step to the establishment of spheres of influence. Asia has never really been the same since.

C China was being carved up. It was in no position to be expanding.

D Had China been stronger and more unified, this is exactly what would have happened. But the superior military power of the Europeans overwhelmed the Chinese.

38 H **Choose the answer choice that wouldn't be a reason that monarchs in the sixteenth century would have accepted Martin Luther's rebellion. If you aren't sure, try to pick the one that would make the most sense.**

F Control of church-owned property and revenues meant a huge increase in the king's assets. Therefore, the monarchs supported this reason, and this answer choice is wrong. Martin Luther and his supporters rebelled against the church's corrupt practices. Without papal authority, the monarch could confiscate church property and suppress taxes levied by the church.

G This is yet another reason for a monarchial alliance with Martin Luther. In addition to the king's regular supporters, with Luther's new church, the king's political power included the newly reformed church, its property, and ministers. (Easier said than done in most areas, because accomplishing this greedy control translated to many bloody years.)

H This answer choice is correct because the monarchy was least interested in participation or control of the citizens. Even if they were interested, the Reformation was not the answer. Kings thrived on wealth and power with the least interference.

J As the popularity of Luther's movement increased, those monarchies favorable toward Martin Luther also gained access to a large network of heretics eager to fight intrusive papal authorities.

39 D Read the quote. If you missed this question, read it again.

A Napoleon tried to sweep across Europe after the French Revolution, but this quote refers to how things are made and what's available to more people.

B The Commercial Revolution involved a switch from a barter economy to a monetary, banking, and investment economy nearly 200 years before·there would have been a quote about "assembly lines."

C Did Russia "sweep across Europe?" Nope. This is about factories and the economy.

D The Industrial Revolution was all about factories. Factories were all about mass production, division of labor, and assembly lines. People stopped working out of their homes (the domestic situation) and went to work in the giant factories that started springing up all over Europe.

40 G Look for an answer choice that describes something that is not characteristic of a traditional, rural community. This question is asking you to identify something that has changed, not something that has stayed the same.

F The authority of religious leaders generally declines in industrial societies because people are exposed to many new things. The overall role of the Church and the absolute authority of religious leaders are often reduced as new ideas, philosophies, and lifestyles compete for people's attention.

G As a nation modernizes, the traditions change, otherwise it wouldn't be called modernization. Specifically, the roles of women often change as nations industrialize and urbanize because city life imposes different expectations on women than rural life. Women are often more educated in industrial societies and look for work that utilizes their education and provides enough income for the expenses of city life. While not all societies change their values of women, the question asks for the answer choice that occurs *most often*. Be aware of those kinds of terms.

H Generally, the family size decrease as nations industrialize because the cost of having children is higher in the cities than in rural areas, and children are not valuable as workers at a young age, as they often are on the farms. In addition, the religious practices in many traditional societies often discourage family planning in favor of large families, so as nations industrialize and religion becomes less influential, the religious reasons for large families also begin to fade.

J The cost of medical care actually increases as nations industrialize because the quality of medical care is superior in industrialized nations. It costs a lot of money for hospitals to be built and medical advances to be brought to the people. They pay a lot more than they had before, but they get treatments and care that they didn't have access before.

41 C **Africa is a big continent. Try to think of just a few things that you know about Africa. The Nile River. The Sahara Desert. The coastlines. Think of a movie you've seen if you have to. Anything that will help you to see Africa.**

A Nope. There are rain forests.

B No.

C Yes! River valleys. Rain forests. Deserts. That's pretty diverse. Africa is known for its geographical diversity.

D Nuh-uh. The Nile is one of the world's largest river systems.

42 G **Take note! The best approach to this question comes from examining the most similar responses associated with (1) religion and (2) specifically, Hinduism. The rest is easy, regardless of your familiarity with Taoism.**

F Reincarnation is indeed a significant component of India's primary religion, Hinduism.

G Taoism, an influential Chinese philosophy, is second in that area to Confucianism. Taoism stresses the importance of living naturally, according to nature's pattern. Its converts reject ambition in favor of a meditative return to nature. Unlike Buddhism which gained a foothold in India, Taoism remains apart from Indian culture.

H Caste is distinctly associated with old and new India, Hinduism specifically. While contemporary Indian caste systems have abandoned some restrictions, many rural areas persist in the social and religious restrictions based on caste.

J The term karma is distinctly associated with ancient Indian civilization as well as Hinduism. Karma refers to the law of moral consequences which holds that one's status in present life has been determined by the deeds of previous lives.

43 C **Think of everything you know about the Middle East during the modern era. Think of Iraq, Iran, Saudi Arabia, Syria, or other countries in the region. Eliminate answer choices that are inconsistent with the region as a whole.**

A The Middle East is not a model of economic and political equality. With the exception of Israel, the nations of the Middle East are all Muslim. The traditions of the religion have prevented broad reforms that otherwise might have occurred. And the recent uprisings of Islamic fundamentalists after the success of the Iranian Revolution in 1979 indicate that many in the region want to eliminate some of the reforms that *have* occurred.

B The European spheres of influence are no longer in existence in the Middle East. When Gamal Abdel-Nasser of Egypt nationalized the Suez Canal, the Europeans were pretty much kicked out of the region entirely.

C Bingo! The Middle East contains roughly two-thirds of the world's petroleum reserves. Any instability in the region sends shock waves throughout the industrialized, gas-guzzling world. Remember the Persian Gulf War?

D No. The region is a primary source of petroleum, not uranium.

44 G First, realize that this question involves the 1930s, a time of worldwide depression. Second, realize that it's asking about the Nazi Party, which you should be somewhat familiar with. Put the two together and use your common sense.

F This treaty officially ended World War I, and it was hated by most Germans, who were humiliated by provisions that severely reduced Germany's ability to develop a strong military or economy. The Nazis advocated ignoring the treaty.

G Of course! It's a time of depression, and it's especially bad in Germany because of the Treaty of Versailles. The people supported the party that promised to bring economic prosperity.

H Absolutely not! If you know only a little about the Nazi Party, you must remember that it was not a multicultural group. They believed their own race to be superior to all others.

J No. The international organizations were against the Nazi Party, mostly because the party advocated ignoring the Treaty of Versailles, and then did not respect the borders of other nations.

45 A If you've been reviewing Hitler in your materials, good for you. You should review the Treaty of Versailles as well, specifically with regard to its impact on Germany. Even if you simply remember that the Treaty of Versailles was signed after World War I, you should be able to put these events in order.

A The Treaty of Versailles was signed at the end of World War I. It blamed Germany for the war and punished it harshly—so harshly that Germany was humiliated and thrown into economic crisis. Years later, Adolf Hitler rose to power by blaming the Treaty of Versailles for the crisis in Germany and inciting the masses to assert themselves as a powerful German nation. After Hitler became the chancellor of Germany (and then the totalitarian *fuhrer*), German troops invaded Poland, marking the beginning of World War II.

B It was Hitler who ordered the invasion of Poland, and he wasn't able to do that until after he became chancellor.

C Adolf Hitler became chancellor of Germany long after the signing of the Treaty of Versailles. In fact, his frustration over the Treaty of Versailles helped to shape his political philosophy, which in turn facilitated his rise to power.

D It was Hitler who ordered the invasion of Poland, and he wasn't able to do that until after he became chancellor.

46 F If you don't know the terms in the answer choices, then isn't it about time that you started reviewing?

F *Zionism* has traditionally referred to the Jewish movement to establish a homeland in Palestine, and ever since the nation of Israel was created, Zionism has referred to the movement to maintain it.

G *Secularism* is a term used to describe indifference to religion. The secular world is the part of the world that doesn't have anything to do with religion. It's not necessarily against religion, it's just apart from religion. Mathematics is secular, for example. The Jewish movement to establish a homeland in Palestine was most definitely religiously motivated, and therefore cannot be described as secular.

H *Animism* is a group of traditional African religions. Animism has nothing to do with the Jewish movement to establish a homeland in Palestine.

J *Marxism* is the economic philosophy at the heart of socialism and communism. It has nothing to do with the Jewish movement to establish a homeland in Palestine. Look up *Karl Marx* in the review materials if you picked this answer.

47 A **You absolutely have to know that after World War II, European colonialism fell because of independence movements in more than 50 African nations. It's essential world history!**

A This is what happened on the continent from top to bottom.

B The new African nations eagerly joined international organizations to increase their voice and visibility in world politics and economics.

C Absolutely not. Most nations have welcomed assistance, especially when it has involved technological innovations.

D In the years immediately following World War II, creating a unified Africa was a goal of some leaders, but most African nations have turned their concerns inward to solving domestic problems that continue to plague their development.

48 J **Choose the answer choice that you just can't argue with. If you're still not sure, start eliminating answer choices that are inconsistent with Soviet, and therefore communist, ideology.**

F The Soviet Union did not support fascism. Think of which side the Soviet Union fought on during the wars. Lenin and Stalin didn't like fascism because fascists didn't like communism.

G Communism as an economic system is opposed to capitalism. Enough said.

H Communism as a political system is opposed to democracy.

J Sure. The only reason to establish satellites (which essentially means establishing miniature models of yourself) is to increase your power and influence.

49 A **If you don't know Gandhi already, get to know him. He's a favorite of the test writers. Gandhi definitely did not like violence.**

A Gandhi gained independence for India from Britain in 1947 by organizing peaceful resistance, such as boycotts and hunger strikes, to British authority. He is one of the world's most famous practitioners of civil disobedience.

B Gandhi was Hindu, not Islamic.

C Gandhi was not opposed to Hindus holding political office. Nor was he opposed to Islamic people holding political office. The Islamic people, however, formed their own nation of Pakistan after independence was won from Britain.

D If you remember nothing else about Gandhi, remember that he was strongly opposed to violence.

50 H **You want to find a geographic feature that not only describes Russia, but also explains its foreign policy and economic development. You probably know that Russia's foreign policy has traditionally been expansionist. Therefore, think of a geographic feature that would explain Russia's desire to expand.**

F If Russia lacked natural resources, this fact would certainly influence Russia's foreign policy and economic development. But Russia is an enormous piece of land with a variety of natural resources including copper, coal, lead, and forests.

G Russia has deserts in the southeast, but they are not vast, and they have not significantly affected foreign policy or economic development.

H This is it. Much of Russia is either land-locked or enclosed by frozen seas to the north. Its economic development has been hindered by this fact. Therefore, its foreign policy has attempted to gain access to warm water ports, especially to the west along the Baltic Sea and to the south along the Black Sea.

J Russia has extensive mountain ranges in the east. However, these areas are sparsely populated and have not significantly affected the foreign policy or economic development of the nation.

51 D **You need to know about apartheid. Notice that answer choice B and answer choice C are direct opposites. When two direct opposites appear, it is likely that one of them is the answer choice.**

A The policy of apartheid did not involve the forced migration of blacks to other countries. It did involve the forced migration of blacks within the nation of South Africa.

B No. This is the opposite of the policy of apartheid. Under apartheid, the races were separated, not integrated.

C Apartheid was not an immigration system but an internal political and cultural policy.

D Yes. The policy of apartheid divided (segregated) the races in South Africa. Blacks were not only segregated from whites but also denied many of the rights that were granted to whites.

52 J **You need to determine what is meant by the word involvement. The 1960s and 1970s was the time of the Vietnam War, and since Vietnam is in Southeast Asia, you should conclude that involvement means military involvement.**

F Gaining new markets is certainly a reason to be involved in a region, but we want an answer choice that explains why the United States was sending troops into Southeast Asia in the 1960s and 1970s.

G New sources of oil were important objectives but not in Southeast Asia.

H The United States was not looking for colonies anywhere.

J Yes. The United States sent troops, weapons, and money to Southeast Asia in an effort to stop the spread of communism to Vietnam. You should be aware that preventing the spread of communism was the major foreign policy goal of the United States from the end of World War II to the end of the Cold War.

53 A **First eliminate any answer choices that just don't make sense. Then try to remember either or both of the codes. Are the codes similar or different?**

A The two codes are similar, requiring warriors to commit themselves to their lords, to their cause, and to fighting bravely till death. Because they developed independently in two different parts of the world, the two codes show that different societies develop similar systems to meet similar needs.

B These two codes show similarities in cultures, not differences.

C Clearly, if force was never used then there would be no need for warriors. Nevertheless, these two codes don't illustrate that nations are trying to conquer *other* nations. The warriors who pledged themselves to these codes were protecting their lords from not only other nations but also rebels within their own kingdom.

D This answer choice just doesn't make sense so get rid of it!

54 G **You need to know only about one of these rebellions in order to answer this question correctly.**

F None of these rebellions were reactions to rapid industrialization. If you're guessing, look for an answer that you can be certain caused rebellions.

G This is the best answer. All three rebellions were motivated by a desire to drive the Europeans out of their colonies. The Sepoy Mutiny in India involved a rebellion against British ships. The Mau Mau in Kenya led very violent revolts against the British in Kenya and won independence within about five years. The Boxer Movement in China was also very violent and involved the murders of European missionaries and mercantilists who had established spheres of influence.

H None of these rebellions were reactions to Mongol domination.

J The Sepoy Rebellion and Boxer Rebellion occurred prior to World War I. The Mau Mau Uprising occurred in Kenya in the 1950s.

55 C **The graph gives us eight years with bar lines above them. These bars split in order to show the number of people who live or are expected to live in developing countries as well as those who live or are expected to live in developed countries. Read each answer choice carefully, compare it to the graph, and eliminate.**

A This can't be true. The lines are getting longer and longer, which means the number of people on earth is getting bigger and bigger.

B In 1970 the world's population was about 3.7 billion people. In 1989 it was 5.2 billion. If the world's population tripled during this time, then there would have been more than 10 billion people in 1989 instead of 5.2 billion.

C In each and every year on the chart, the black line (which represents the number of people in developing countries) is longer than the gray line (which represents the number of people in developed countries). Therefore, most of the population lives in economically developing countries.

D The chart gives us no information regarding who consumes what or where they consume it.

56 F **You're not going to be able to answer this question by looking at the graph, but the graph DOES tell you that the population is growing faster in the developing countries than it is in the countries that are already developed. Now try to think of why this might be true.**

F If the people in developed countries are planning the number of kids they want to have, while the people of developing countries are having a lot of children without any planning, then this would help to explain why the population is growing faster in developing countries. Birth control and other types of contraception are much more readily available in developed nations, as well.

G While pollution is a problem in both developed and developing countries, there is no evidence that it has significantly affected birth rates in either type of country.

H The breakdown of extended families has occurred in developed countries and is beginning to occur in developing countries, where it might lead to fewer births since the extended family won't be around to raise the small children. However, at this point in time, the breakdown of families in *developed* countries contributes to the difference in growth rates.

J In both developing and developed countries, the number of children born into households with only one parent has increased. Therefore, it doesn't explain a significant difference between the two types of countries.

57 B **Following World War II, it was determined that Germany's aggression had been the cause for the war. The Versailles treaty allowed for Germany to pay back its debts.**

A This statement is inconsistent with the question. The United States was a victor; therefore, recruitment was not a priority at this time.

B Yes. This statement was the Allies' justification for reparations payments from the defeated Central powers, particularly Germany.

C Nope, England and the United States remained allies throughout several wars.

D Italy and France waged a tariff war, but this was unassociated with the statement in the question. This can't be the right choice then.

58 H The question asks you to find the best evidence that Spain (as opposed to some other country) was the dominant colonial power in Latin America. In order to do this, you should focus on something that is (or was) true of Spain and is true of Latin America. In other words, pick an answer choice that describes something in common between the two regions of the world. In that way, we would have evidence that the two regions might be connected in some way.

F Spain's monetary unit is the *peseta*. Mexico's is the *peso*. Even if you haven't devoted your life to the memorization of monetary units, you should still raise your eyebrows at this answer choice. Spain and Mexico are sovereign nations, after all. Traditionally, sovereign nations have had their own currencies. Interestingly, developments in modern Europe are changing this, since most of the European Union is moving toward the use of a common currency while retaining sovereignty, but this is a radical departure from the past.

G Spain does not continue to provide military support for Latin America and has not been militarily involved in Latin America since the various countries of the region gained independence.

H In Spain, the official language is Spanish (no surprise there). And in Latin America, most people speak Spanish, except in Brazil where most people speak Portuguese. How is it that Spanish came to be the dominant language of Latin America? Easy. Because Spain was the dominant colonial power in Latin America.

J Argentina and Spain are sovereign countries. The citizens of one sovereign country do not elect representatives to the legislatures of other sovereign countries. Only citizens of Spain elect representatives to the Spanish parliaments, known as the *Cortes*.

59 D Use your common sense on this one and focus on the individual words in the answer choices. If you want people's lives to improve, which of the ideas in the four answer choices should you implement?

A "Borrowing" is a bad word if you're trying to improve your life. Sometimes you have to borrow in order to improve (school loans, for example), but isn't there a better answer choice? Plus, if you're going to borrow money, borrow it to buy stuff that will help your country long-term since, most likely, it's going to take you a long time to repay the loan. Food is obviously essential, but wouldn't it be better to borrow money for farm equipment or irrigation systems, for example, so that you could grow your own food year after year? Borrowing money to buy food that will be gone as soon as it's eaten only creates a cycle wherein you need to borrow more money to buy more food year after year. In times of crisis, some countries have to do this, but it's not the best way to help developing nations improve their standard of living.

B "Relying on a single cash crop" isn't nearly as good an idea as "relying on a lot of cash crops." This answer choice just doesn't make any sense. Why would a country intentionally rely on just one crop? What happens if world demand for that crop decreases? What happens if some other country produces the same crop more efficiently and can sell it for less? What happens if a drought hits? Or a big swarm of locusts? Or what happens if you deplete the soil of its nutrients because you keep growing the same crop on it? All of these are potential nightmare problems for countries that rely on just a single cash crop.

C The presence of a *trade deficit* means you're buying more than you're selling. In other words, you're losing money. If you want to help a nation improve its standard of living, you'd want it to *gain* money, not lose it. Encouraging an increase in the trade deficit would probably be a bad idea, then, don't you think?

D This is by far the best answer choice. First, it has that wonderful word when it comes to improving a nation's economy: *investing*. What's more, it doesn't suggest that the country invest in merely anything, but in "the development of human resources," which more simply means stuff like education and job training. If a country invests in its people power, it has a much better chance at succeeding long-term and in a whole variety of ways beyond just economically.

60 F **Focus on the word strategic. Even if you don't know anything about any of these waterways, you should be able to eliminate two of the answer choices just by focusing on this word. Even if you know only about one of these waterways, you should be able to answer this question correctly.**

F This is the best answer. They are *strategically* important because they give power to the nations who control these waterways. The Suez Canal links the Mediterranean Sea to the Red Sea, saving ships traveling between Europe and Asia the thousands of miles that they otherwise would travel around the tip of Africa. Egypt, which controls the Suez Canal, makes a tremendous amount of money by charging fees to use the canal. Located in Turkey, the Dardenelles and the Bosphorus connect the Black Sea to the Aegean Sea, which is connected to the Mediterranean Sea. Since Russia relies on its Black Sea ports, Turkey is the economic beneficiary of goods that must travel through the Dardenelles and the Bosphorus en route to and from Russia. Finally, the Strait of Hormuz sits at the mouth of the Persian Gulf and is, therefore, the waterway through which the Persian Gulf's oil is distributed to the rest of the world.

G The Suez Canal isn't a natural boundary in the first place—it's man-made, and it's not a boundary. The other waterways are indeed natural, but they don't separate warring nations. They have played roles in wars, to be sure, but they're most significant strategic importance, especially in recent decades, has been economic.

H None of these waterways are along the Tropic of Cancer, the equator, or the Tropic of Capricorn. Even if they were, it wouldn't explain why these waterways are strategically important.

J All of these waterways consist of salt water, not fresh water.

61 D **We want something geographic that affects Southeast Asia. Think of Hong Kong. Think of Singapore. Think of Korea.**

A Definitely not. China isn't democratic.

B Southeast Asia has no large deposits of coal and diamonds, but even if it did, the native population would work the mines.

C There aren't any big deserts in Southeast Asia.

D Waterways. Waterways. Everywhere there are waterways. Look at a map. It's pretty incredible.

62 H Know your geography. If you do, this question practically answers itself. If you don't, you should know that Poland has been attacked a lot, so try to think of a geographic feature that would allow it to be easily attacked.

F Poland probably doesn't have what would be described as a "severe" climate. It's warm in the summer and cool or cold in the winter. This type of climate probably wouldn't have a major influence on its historical and cultural development. You want an answer choice that you can point to and say, "Hey, that factor has influenced things!"

G If Poland had vast deposits of oil, then those deposits would certainly have influenced the country's development, but it would have been its economic development. However, Poland doesn't have vast oil deposits, and we want an answer choice that influences history and culture.

H Yes! Poland is on a plain. You should definitely be aware that if a nation is a farming nation, its culture is significantly affected by that fact. Also, because of the lack of mountains or other natural barriers, Poland has been easily attacked by its neighbors. Therefore, this fact has influenced Poland's history and culture.

J Poland doesn't have a rugged coastline.

63 D Why do people revolt? They revolt because they are unhappy, not because they are happy. Choose the answer choice that describes people who are unhappy.

A If they're prosperous, why would they want to revolt?

B If they're able to climb the ladder, why would they revolt?

C If they can vote and affect the political decisions of the nation, why would they revolt?

D If they're dissatisfied, they might revolt. This answer choice just plain makes good sense.

64 G If you're clueless about this time period, review your materials.

F No. The Mongols weren't big fans of the arts.

G Yes! The Mongols kept Russia more isolated than it was before or after.

H No. Industrialization hadn't yet occurred anywhere during the time of the Mongul rulers.

J No. Democratic reforms hadn't occurred anywhere yet, and they were not encouraged in Russian society until recently.

65 C If you know the approximate time period of the Protestant Reformation, you should be able to eliminate a few answers. Also, you should be aware that the Protestant Reformation was a theological and intellectual movement, carried forth with writings and words, not with guns.

A The steam engine was not yet invented during the Protestant Reformation. The steam engine was fundamental to the Industrial Revolution.

B The light bulb was not yet invented during the Protestant Reformation.

C The Protestant Reformation owes part of its success to the printing press. It allowed Martin Luther's writings to be distributed quickly to the masses. What's more, Luther argued that the Bible should be printed in German so that ordinary people could read the Scriptures rather than rely solely on the Church's interpretation; the printing press made that goal a reality.

D Gunpowder was developed centuries before the Protestant Reformation in China. In many ways, gunpowder has changed the world, but the Protestant Reformation was not primarily a military conflict.

66 J **The Native American population in Mexico dropped by more than 23 million people between 1492 and 1608. Eliminate answer choices that probably wouldn't account for such a dramatic decrease in the population.**

F Poor weather conditions sometimes hurt crops prior to 1492 as well as after, and yet the Native American population managed to grow to 25 million in 1492! If the crop failures were so significant that so many deaths resulted, then the Spanish who were colonizing Latin America would probably have died as well since they needed to eat too.

G Native Americans did not emigrate to Europe and Africa. America was their home. They would have been perfectly happy if the Europeans never came.

H Native groups went to war against Spain, but they mostly stopped fighting with each other after the Spanish arrived because they couldn't fight each other and Spain as well.

J This is the best answer. The Spanish brought with them diseases from Europe that the Native American population had never been exposed to. Because of the lack of exposure, they hadn't built up immunity against the diseases and were wiped out by the millions.

67 B **During the Age of Absolutism, European monarchs attempted to gain absolute control over their territories by taking power away from the local feudal lords and centralizing it. They believed that no one could or should challenge their authority over their territories, even to the point of claiming that they had a divine right under God to rule their lands.**

A The European monarchs during this time period were interested not in human rights, but in their own rights of ruling the citizenry.

B The more centralized the power, the less likely that there will be any attempts against authority. This allowed the monarchs to rule *absolutely*.

C Many European monarchs actually fought Muslim rulers during this time period, so they certainly weren't being neighborly.

D Cooperative farms didn't come along until after the Age of Absolutism.

68 F **When Germany and Italy unified, they became more powerful. Which answer choice describes what happens when two nations on a continent of many nations gain power?**

F The unification of Germany and Italy upset the balance of power in Europe. Suddenly, Germany and Italy were much more powerful than they had been. When Germany, Italy, and Austria-Hungary joined forces as the Triple Alliance, the balance of power was dramatically tilted in their favor, so Russia, Great Britain, and France created the Triple Entente in an attempt to balance the power once again.

G The unification of Germany and Italy had nothing to do with an increase in trade with Russia.

H Absolutely not. When nations unify, there is a tremendous feeling of pride and nationhood. When nations divide into smaller nations, there is also a tremendous feeling of pride and nationhood. Whenever a nation acquires new borders and embarks on a new path, nationalism grows.

J No. As Germany and Italy unified, their neighbors became weaker. Even when the continent was relatively peaceful, the weaker neighbors worried about the power of Italy and Germany, and so tensions increased. Finally, the alliance systems of the Triple Alliance and the Triple Entente developed.

69 D **You should be aware that the Soviet Union was a one-party communist nation that was extremely centralized and government-controlled. You should also know that under Stalin, the nation's economy was focused on the build-up of heavy industry.**

A No. The Soviet Union has never been a nation that has experienced an abundance of consumer goods. The government owned the means of production and determined how much of what kinds of goods would be available (a command economy). Since the government was focused on the build-up of heavy industry, it did not provide a surplus of consumer goods.

B The Soviet Union was remarkably stable under Stalin, most probably because he was tremendously powerful and tremendously feared. He ruled absolutely, so few insurrections occurred. If they did, they were dealt with swiftly and often brutally.

C This is entirely inconsistent with communism in general and the Soviet Union in particular. Stalin collectivized the farms, which means that the government took ownership of the farmlands and directed the farm workers.

D This is the best answer. Stalin's Soviet Union did not tolerate insurrection or alternative ideas. The secret police was used to quiet any insurrections or those suspected of having opinions contrary to those of the government.

70 H **This question asks you to determine why two countries, each with a lot of bombs, wouldn't want to get in a fight. As long as you know that both countries have more nuclear weapons than this test has questions, you should choose the answer that is hardest to argue with.**

F Both the former Soviet Union and the United States are and were more powerful than the United Nations, and both nations have veto power within the United Nations Security Council. So while no one wanted the Cold War to become a hot war, the United Nations wouldn't have been able to keep them from getting into a war if that's what the two sides really wanted to do. But they didn't want to. Why was that?

G The Soviet Union and the United States were the two most powerful nations on Earth during the Cold War. Clearly, many nonaligned nations didn't want to see a war develop and worked to keep the peace, but nonaligned nations were generally not in a position to pressure these two superpowers.

H This is the best answer because it is hard to argue with. Both sides have enough bombs to destroy the world, and if the globe is destroyed, then there's no use fighting because there won't be anything worth winning or anybody left to say who's won.

J The former Soviet Union and the United States actually fed the tensions in the Middle East by providing opposing armies with the weapons to fight. Therefore, the tensions in the Middle East arguably increased the chances of global destruction.

71 D **You want to choose an answer choice that sounds like a "major goal" as opposed to an answer choice that is too detailed or too narrow.**

A One-product economies are risky. The European Community supports diversification.

B This answer choice doesn't sound like a "major" goal, but rather just a small part of a larger plan. The European Community has forgiven some debts, but there's a lot more to the European Community than just this small fact.

C The European Community can't repay the loans because no loans were granted in the first place. Under the Marshall Plan, the United States *gave* money to Europe (grants), so no money needs to be repaid.

D Now this sounds like a major goal! One of the major goals of the European Community is to utilize the combined economic strength of many European nations to compete with the economies of Japan and the United States.

72 F **You need to know about only one of these groups in order to answer this question correctly.**

F This is the best answer. All three groups tried to drive the Europeans out of their countries. All three groups also used violence in their attempts. The Sepoy Mutiny in India involved a rebellion against British ships. The Mau Mau in Kenya led very violent revolts against the British, and won independence within about five years. The Boxer Movement in China was also violent and involved the murders of European missionaries and mercantilists who had established spheres of influence.

G No. These movements were against Western aggression, not supported by it.

H While Marxism motivated some groups in Africa, Marxism wasn't a factor in India during the Sepoy Mutiny or in China during the Boxer Movement.

J No. All three were groups used violence.

73 D **A primary source is one that gives a first-hand account of the event that you're researching. If you want to know the details about an event, it's better to ask someone who was actually there!**

A A textbook on Russian history isn't a primary source. Even if the author of the textbook is Russian, it's impossible that he or she was personally at every event discussed in the textbook. Russia's history spans thousands of years and thousands of miles!

B Even if the encyclopedia article were written by a practitioner of a religion in the Middle East, it's unlikely that the article is about personal experiences, and even if it were, the writer is likely not a Jew, Christian, and Muslim simultaneously. Therefore, the article would contain a lot of second-hand information.

C A novel is not a primary source. It's an imagined story!

D A diary is a primary source. The diary would likely include the writer's personal, first-hand accounts of experiences within the concentration camp.

Virginia SOL: EOC World History & Geography Scoring Guide:

Practice Test 1: World History to AD 1000 & Geography

Total 71 questions	__/71 correct
– 10 Field-test items	
(3, 11, 15, 21, 27, 34, 45, 51, 58, 70)	__/10 correct

Total 61 questions	__/61 correct
	0–32 = failing
	33–54 = passing
	55–61 = pass/advanced

Practice Test 2: World History from AD 1000 to the Present & Geography

Total 73 questions	__/73 correct
– 10 Field-test items	
(4, 15, 23, 26, 31, 33, 43, 50, 56, 69)	__/63 correct

Total 63 questions	__/63 correct
	0–35 = failing
	36–56 = passing
	57–63 = pass/advanced

Completely darken bubbles with a No. 2 pencil. If you make a mistake, be sure to erase mark completely. Erase all stray marks.

1. YOUR NAME: _____
(Print)　　　　　Last　　　　　　　　First　　　　　　　　M.I.

SIGNATURE: _____　　**DATE:** _____ / _____ / _____

HOME ADDRESS: _____
(Print)　　　　　　　　　　　　　Number

_____
City　　　　　　　State　　　　　　　Zip Code

PHONE NO.: _____
(Print)

IMPORTANT: Please fill in these boxes exactly as shown on the back cover of your test book.

2. TEST FORM

3. TEST CODE　　　**4. REGISTRATION NUMBER**

6. DATE OF BIRTH

Month	Day	Year
◯ JAN		
◯ FEB		
◯ MAR	⓪ ⓪	⓪ ⓪
◯ APR	① ①	① ①
◯ MAY	② ②	② ②
◯ JUN	③ ③	③ ③
◯ JUL	④	④ ④
◯ AUG	⑤	⑤ ⑤
◯ SEP	⑦	⑦ ⑦
◯ OCT	⑧	⑧ ⑧
◯ NOV	⑨	⑨ ⑨
◯ DEC		

7. SEX
◯ MALE
◯ FEMALE

The Princeton Review
© 1996 Princeton Review L.L.C.
FORM NO. 00001-PR

5. YOUR NAME

First 4 letters of last name　FIRST INIT　MID INIT

(bubble columns A–Z)

Practice Test ①

1. Ⓐ Ⓑ Ⓒ Ⓓ
2. Ⓕ Ⓖ Ⓗ Ⓙ
3. Ⓐ Ⓑ Ⓒ Ⓓ
4. Ⓕ Ⓖ Ⓗ Ⓙ
5. Ⓐ Ⓑ Ⓒ Ⓓ
6. Ⓕ Ⓖ Ⓗ Ⓙ
7. Ⓐ Ⓑ Ⓒ Ⓓ
8. Ⓕ Ⓖ Ⓗ Ⓙ
9. Ⓐ Ⓑ Ⓒ Ⓓ
10. Ⓕ Ⓖ Ⓗ Ⓙ
11. Ⓐ Ⓑ Ⓒ Ⓓ
12. Ⓕ Ⓖ Ⓗ Ⓙ
13. Ⓐ Ⓑ Ⓒ Ⓓ
14. Ⓕ Ⓖ Ⓗ Ⓙ

15. Ⓐ Ⓑ Ⓒ Ⓓ
16. Ⓕ Ⓖ Ⓗ Ⓙ
17. Ⓐ Ⓑ Ⓒ Ⓓ
18. Ⓕ Ⓖ Ⓗ Ⓙ
19. Ⓐ Ⓑ Ⓒ Ⓓ
20. Ⓕ Ⓖ Ⓗ Ⓙ
21. Ⓐ Ⓑ Ⓒ Ⓓ
22. Ⓕ Ⓖ Ⓗ Ⓙ
23. Ⓐ Ⓑ Ⓒ Ⓓ
24. Ⓕ Ⓖ Ⓗ Ⓙ
25. Ⓐ Ⓑ Ⓒ Ⓓ
26. Ⓕ Ⓖ Ⓗ Ⓙ
27. Ⓐ Ⓑ Ⓒ Ⓓ
28. Ⓕ Ⓖ Ⓗ Ⓙ

29. Ⓐ Ⓑ Ⓒ Ⓓ
30. Ⓕ Ⓖ Ⓗ Ⓙ
31. Ⓐ Ⓑ Ⓒ Ⓓ
32. Ⓕ Ⓖ Ⓗ Ⓙ
33. Ⓐ Ⓑ Ⓒ Ⓓ
34. Ⓕ Ⓖ Ⓗ Ⓙ
35. Ⓐ Ⓑ Ⓒ Ⓓ
36. Ⓕ Ⓖ Ⓗ Ⓙ
37. Ⓐ Ⓑ Ⓒ Ⓓ
38. Ⓕ Ⓖ Ⓗ Ⓙ
39. Ⓐ Ⓑ Ⓒ Ⓓ
40. Ⓕ Ⓖ Ⓗ Ⓙ
41. Ⓐ Ⓑ Ⓒ Ⓓ
42. Ⓕ Ⓖ Ⓗ Ⓙ

43. Ⓐ Ⓑ Ⓒ Ⓓ
44. Ⓕ Ⓖ Ⓗ Ⓙ
45. Ⓐ Ⓑ Ⓒ Ⓓ
46. Ⓕ Ⓖ Ⓗ Ⓙ
47. Ⓐ Ⓑ Ⓒ Ⓓ
48. Ⓕ Ⓖ Ⓗ Ⓙ
49. Ⓐ Ⓑ Ⓒ Ⓓ
50. Ⓕ Ⓖ Ⓗ Ⓙ
51. Ⓐ Ⓑ Ⓒ Ⓓ
52. Ⓕ Ⓖ Ⓗ Ⓙ
53. Ⓐ Ⓑ Ⓒ Ⓓ
54. Ⓕ Ⓖ Ⓗ Ⓙ
55. Ⓐ Ⓑ Ⓒ Ⓓ
56. Ⓕ Ⓖ Ⓗ Ⓙ

Completely darken bubbles with a No. 2 pencil. If you make a mistake, be sure to erase mark completely. Erase all stray marks.

1. YOUR NAME: _____
(Print)   Last   First   M.I.

SIGNATURE: _____   **DATE:** ___ / ___ / ___

HOME ADDRESS: _____
(Print)   Number

_____
City   State   Zip Code

PHONE NO.: _____
(Print)

IMPORTANT: Please fill in these boxes exactly as shown on the back cover of your test book.

2. TEST FORM

6. DATE OF BIRTH

Month	Day		Year	
○ JAN				
○ FEB				
○ MAR	⓪	⓪	⓪	⓪
○ APR	①	①	①	①
○ MAY	②	②	②	②
○ JUN	③	③	③	③
○ JUL		④	④	④
○ AUG		⑤	⑤	⑤
○ SEP		⑦	⑦	⑦
○ OCT		⑧	⑧	⑧
○ NOV		⑨	⑨	⑨
○ DEC				

3. TEST CODE   **4. REGISTRATION NUMBER**

⓪	Ⓐ	⓪	⓪	⓪	⓪	⓪	⓪	⓪	⓪	⓪
①	Ⓑ	①	①	①	①	①	①	①	①	①
②	Ⓒ	②	②	②	②	②	②	②	②	②
③	Ⓓ	③	③	③	③	③	③	③	③	③
④	Ⓔ	④	④	④	④	④	④	④	④	④
⑤	Ⓕ	⑤	⑤	⑤	⑤	⑤	⑤	⑤	⑤	⑤
⑦	Ⓖ	⑦	⑦	⑦	⑦	⑦	⑦	⑦	⑦	⑦
⑧		⑧	⑧	⑧	⑧	⑧	⑧	⑧	⑧	⑧
⑨		⑨	⑨	⑨	⑨	⑨	⑨	⑨	⑨	⑨

7. SEX
○ MALE
○ FEMALE

The Princeton Review
© 1996 Princeton Review L.L.C.
FORM NO. 00001-PR

5. YOUR NAME

First 4 letters of last name				FIRST INIT	MID INIT
Ⓐ	Ⓐ	Ⓐ	Ⓐ	Ⓐ	Ⓐ
Ⓑ	Ⓑ	Ⓑ	Ⓑ	Ⓑ	Ⓑ
Ⓒ	Ⓒ	Ⓒ	Ⓒ	Ⓒ	Ⓒ
Ⓓ	Ⓓ	Ⓓ	Ⓓ	Ⓓ	Ⓓ
Ⓔ	Ⓔ	Ⓔ	Ⓔ	Ⓔ	Ⓔ
Ⓕ	Ⓕ	Ⓕ	Ⓕ	Ⓕ	Ⓕ
Ⓖ	Ⓖ	Ⓖ	Ⓖ	Ⓖ	Ⓖ
Ⓗ	Ⓗ	Ⓗ	Ⓗ	Ⓗ	Ⓗ
Ⓘ	Ⓘ	Ⓘ	Ⓘ	Ⓘ	Ⓘ
Ⓙ	Ⓙ	Ⓙ	Ⓙ	Ⓙ	Ⓙ
Ⓚ	Ⓚ	Ⓚ	Ⓚ	Ⓚ	Ⓚ
Ⓛ	Ⓛ	Ⓛ	Ⓛ	Ⓛ	Ⓛ
Ⓜ	Ⓜ	Ⓜ	Ⓜ	Ⓜ	Ⓜ
Ⓝ	Ⓝ	Ⓝ	Ⓝ	Ⓝ	Ⓝ
Ⓞ	Ⓞ	Ⓞ	Ⓞ	Ⓞ	Ⓞ
Ⓟ	Ⓟ	Ⓟ	Ⓟ	Ⓟ	Ⓟ
Ⓠ	Ⓠ	Ⓠ	Ⓠ	Ⓠ	Ⓠ
Ⓡ	Ⓡ	Ⓡ	Ⓡ	Ⓡ	Ⓡ
Ⓢ	Ⓢ	Ⓢ	Ⓢ	Ⓢ	Ⓢ
Ⓣ	Ⓣ	Ⓣ	Ⓣ	Ⓣ	Ⓣ
Ⓤ	Ⓤ	Ⓤ	Ⓤ	Ⓤ	Ⓤ
Ⓥ	Ⓥ	Ⓥ	Ⓥ	Ⓥ	Ⓥ
Ⓦ	Ⓦ	Ⓦ	Ⓦ	Ⓦ	Ⓦ
Ⓧ	Ⓧ	Ⓧ	Ⓧ	Ⓧ	Ⓧ
Ⓨ	Ⓨ	Ⓨ	Ⓨ	Ⓨ	Ⓨ
Ⓩ	Ⓩ	Ⓩ	Ⓩ	Ⓩ	Ⓩ

Practice Test ①

57. Ⓐ Ⓑ Ⓒ Ⓓ   71. Ⓐ Ⓑ Ⓒ Ⓓ
58. Ⓕ Ⓖ Ⓗ Ⓙ   72. Ⓕ Ⓖ Ⓗ Ⓙ
59. Ⓐ Ⓑ Ⓒ Ⓓ   73. Ⓐ Ⓑ Ⓒ Ⓓ
60. Ⓕ Ⓖ Ⓗ Ⓙ   74. Ⓕ Ⓖ Ⓗ Ⓙ
61. Ⓐ Ⓑ Ⓒ Ⓓ   75. Ⓐ Ⓑ Ⓒ Ⓓ
62. Ⓕ Ⓖ Ⓗ Ⓙ
63. Ⓐ Ⓑ Ⓒ Ⓓ
64. Ⓕ Ⓖ Ⓗ Ⓙ
65. Ⓐ Ⓑ Ⓒ Ⓓ
66. Ⓕ Ⓖ Ⓗ Ⓙ
67. Ⓐ Ⓑ Ⓒ Ⓓ
68. Ⓕ Ⓖ Ⓗ Ⓙ
69. Ⓐ Ⓑ Ⓒ Ⓓ
70. Ⓕ Ⓖ Ⓗ Ⓙ

Completely darken bubbles with a No. 2 pencil. If you make a mistake, be sure to erase mark completely. Erase all stray marks.

1. YOUR NAME:

(Print) Last First M.I.

SIGNATURE: _____ DATE: ____ / ____ / ____

HOME ADDRESS: _____
(Print) Number

_____
City State Zip Code

PHONE NO.: _____
(Print)

IMPORTANT: Please fill in these boxes exactly as shown on the back cover of your test book.

2. TEST FORM

3. TEST CODE

4. REGISTRATION NUMBER

5. YOUR NAME

First 4 letters of last name				FIRST INIT	MID INIT
A	A	A	A	A	A
B	B	B	B	B	B
C	C	C	C	C	C
D	D	D	D	D	D
E	E	E	E	E	E
F	F	F	F	F	F
G	G	G	G	G	G
H	H	H	H	H	H
I	I	I	I	I	I
J	J	J	J	J	J
K	K	K	K	K	K
L	L	L	L	L	L
M	M	M	M	M	M
N	N	N	N	N	N
O	O	O	O	O	O
P	P	P	P	P	P
Q	Q	Q	Q	Q	Q
R	R	R	R	R	R
S	S	S	S	S	S
T	T	T	T	T	T
U	U	U	U	U	U
V	V	V	V	V	V
W	W	W	W	W	W
X	X	X	X	X	X
Y	Y	Y	Y	Y	Y
Z	Z	Z	Z	Z	Z

6. DATE OF BIRTH

Month	Day	Year
JAN		
FEB		
MAR	0 0	0 0
APR	1 1	1 1
MAY	2 2	2 2
JUN	3 3	3 3
JUL	4	4 4
AUG	5	5 5
SEP	7	7 7
OCT	8	8 8
NOV	9	9 9
DEC		

Test Code bubbles: 0 A, 1 B, 2 C, 3 D, 4 E, 5 F, 7 G, 8, 9

Registration Number bubbles: 0 1 2 3 4 5 7 8 9 (in columns)

7. SEX
MALE
FEMALE

The Princeton Review
© 1996 Princeton Review L.L.C.
FORM NO. 00001-PR

Practice Test ②

1. A B C D
2. F G H J
3. A B C D
4. F G H J
5. A B C D
6. F G H J
7. A B C D
8. F G H J
9. A B C D
10. F G H J
11. A B C D
12. F G H J
13. A B C D
14. F G H J

15. A B C D
16. F G H J
17. A B C D
18. F G H J
19. A B C D
20. F G H J
21. A B C D
22. F G H J
23. A B C D
24. F G H J
25. A B C D
26. F G H J
27. A B C D
28. F G H J

29. A B C D
30. F G H J
31. A B C D
32. F G H J
33. A B C D
34. F G H J
35. A B C D
36. F G H J
37. A B C D
38. F G H J
39. A B C D
40. F G H J
41. A B C D
42. F G H J

43. A B C D
44. F G H J
45. A B C D
46. F G H J
47. A B C D
48. F G H J
49. A B C D
50. F G H J
51. A B C D
52. F G H J
53. A B C D
54. F G H J
55. A B C D
56. F G H J

Completely darken bubbles with a No. 2 pencil. If you make a mistake, be sure to erase mark completely. Erase all stray marks.

1. YOUR NAME:

YOUR NAME: _____
(Print)   Last   First   M.I.

SIGNATURE: _____   DATE: _____ / ___ / ___

HOME ADDRESS: _____
(Print)   Number

_____
City   State   Zip Code

PHONE NO.: _____
(Print)

IMPORTANT: Please fill in these boxes exactly as shown on the back cover of your test book.

2. TEST FORM

3. TEST CODE

4. REGISTRATION NUMBER

5. YOUR NAME

First 4 letters of last name				FIRST INIT	MID INIT

(Bubbles A–Z for each column)

6. DATE OF BIRTH

Month	Day	Year
JAN		
FEB		
MAR	0 0 0 0	
APR	1 1 1 1	
MAY	2 2 2 2	
JUN	3 3 3 3	
JUL	4 4 4	
AUG	5 5 5	
SEP	7 7 7	
OCT	8 8 8	
NOV	9 9 9	
DEC		

7. SEX

○ MALE
○ FEMALE

The Princeton Review
© 1996 Princeton Review L.L.C.
FORM NO. 00001-PR

Practice Test 2

57. Ⓐ Ⓑ Ⓒ Ⓓ
58. Ⓕ Ⓖ Ⓗ Ⓙ
59. Ⓐ Ⓑ Ⓒ Ⓓ
60. Ⓕ Ⓖ Ⓗ Ⓙ
61. Ⓐ Ⓑ Ⓒ Ⓓ
62. Ⓕ Ⓖ Ⓗ Ⓙ
63. Ⓐ Ⓑ Ⓒ Ⓓ
64. Ⓕ Ⓖ Ⓗ Ⓙ
65. Ⓐ Ⓑ Ⓒ Ⓓ
66. Ⓕ Ⓖ Ⓗ Ⓙ
67. Ⓐ Ⓑ Ⓒ Ⓓ
68. Ⓕ Ⓖ Ⓗ Ⓙ
69. Ⓐ Ⓑ Ⓒ Ⓓ
70. Ⓕ Ⓖ Ⓗ Ⓙ

71. Ⓐ Ⓑ Ⓒ Ⓓ
72. Ⓕ Ⓖ Ⓗ Ⓙ
73. Ⓐ Ⓑ Ⓒ Ⓓ
74. Ⓕ Ⓖ Ⓗ Ⓙ
75. Ⓐ Ⓑ Ⓒ Ⓓ

About the Author

Dave Daniel has been a teacher, trainer, and writer for The Princeton Review for eight years. He earned his bachelor's degree from the University of Texas at Austin in 1990 and his J.D. from The Ohio State University College of Law in 1994. He currently lives in Kansas City, Missouri, where he is Director of Counseling Services for The Princeton Review's partnership with Edison Schools. In addition, Dave continues to teach SAT, ACT, GRE, GMAT, and LSAT courses to scores of students every year.

From The Princeton Review

Better Students, Better Scores, Better Schools

Designed to improve standardized test scores.

Parents

- Stay involved with child's classwork and test performance

- Access resources to use at home to help child succeed in school

- Spend quality time with child while directly affecting test scores

Students

- Diagnose which skills are strong and which need improvement

- Work with Homeroom's tailored resources to master each and every weak skill

- Work at their own pace, on their own level

Educators

- Keep track of whole class and individual student progress

- Individualize students' learning

- Maximize school's technology investment

For Math and Reading in Grades 3–8, Homeroom covers:

- CTBS/TerraNova
- ITBS
- SAT-9
- **VA: SOL**
- FL: FCAT
- TX: TAAS
- NY: ELA and Math

For more information:
- Visit www.homeroom.com
- Call 1877-8Homeroom or
- E-mail info@homeroom.com

FIND US...

International

Hong Kong
4/F Sun Hung Kai Centre
30 Harbour Road, Wan Chai,
Hong Kong
Tel: (011)85-2-517-3016

Japan
Fuji Building 40, 15-14
Sakuragaokacho, Shibuya Ku,
Tokyo 150, Japan
Tel: (011)81-3-3463-1343

Korea
Tae Young Bldg, 944-24,
Daechi- Dong, Kangnam-Ku
The Princeton Review- ANC
Seoul, Korea 135-280,
South Korea
Tel: (011)82-2-554-7763

Mexico City
PR Mex S De RL De Cv
Guanajuato 228 Col. Roma
06700 Mexico D.F., Mexico
Tel: 525-564-9468

Montreal
666 Sherbrooke St.
West, Suite 202
Montreal, QC H3A 1E7 Canada
Tel: (514) 499-0870

Pakistan
1 Bawa Park - 90 Upper Mall
Lahore, Pakistan
Tel: (011)92-42-571-2315

Spain
Pza. Castilla, 3 - 5° A, 28046
Madrid, Spain
Tel: (011)341-323-4212

Taiwan
155 Chung Hsiao East Road
Section 4 - 4th Floor,
Taipei R.O.C., Taiwan
Tel: (011)886-2-751-1243

Thailand
Building One, 99 Wireless Road
Bangkok, Thailand 10330
Tel: (662) 256-7080

Toronto
1240 Bay Street, Suite 300
Toronto M5R 2A7 Canada
Tel: (800) 495-7737
Tel: (716) 839-4391

locations

National (U.S.)
We have over 60 offices around the United States and
run courses in over 400 sites. For courses and locations
within the U.S. call 1 (800) 2/Review and you will be
routed to the nearest office.

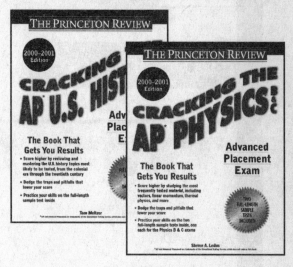

www.review.com

Expert Advice

Talk About It

www.review.com

Pop Surveys

Paying for it

www.review.com

www.review.com

The Princeton Review

Getting in

Word du Jour

www.review.com

Find-O-Rama School & Career Search

www.review.com

Best Schools

Finding it